W9-BFS-621

Technology Integration

The Management of Innovation and Change Series

Michael L. Tushman and Andrew H. Van de Ven, Series Editors

Emerging Patterns of Innovation:
Sources of Japan's Technological Edge
Fumio Kodama, with a Foreword by Lewis M. Branscomb

Crisis & Renewal:
Meeting the Challenge of Organizational Change
David K. Hurst

Winning through Innovation:
A Practical Guide to Leading Organizational Change and Renewal
Michael L. Tushman and Charles A. O'Reilly III

Imitation to Innovation:
The Dynamics of Korea's Technological Learning
Linsu Kim

The Innovator's Dilemma:
When New Technologies Cause Great Firms to Fail
Clayton M. Christensen

Technology Integration:
Making Critical Choices in a Dynamic World
Marco Iansiti

Technology Integration

making critical choices in a dynamic world

Marco Iansiti

Harvard Business School Press
Boston, Massachusetts

Printed in the United States of America

02 01 00 99 98   5 4 3 2 1

Library of Congress Cataloging-in-Publication Data

Iansiti, Marco, 1961–
Technology integration : making critical choices in a dynamic world / Marco Iansiti.
p. cm. — (The management of innovation and change series)
Includes bibliographical references and index.
ISBN 0-87584-787-0 (alk. paper)
1. Research, Industrial—Management. 2. Technology—Management. 3. Computer industry—Management. I. Title. II. Series.
T175.5.I18 1997
607'.2—DC21   97–13996
CIP

The paper used in this publication meets the requirements of the American National Standard for Permanence of Paper for Printed Library Materials Z39.49-1984.

To my father

Contents

Preface *ix*

1 Introduction 1

2 The Concept of Technology Integration 7

3 Studying Technology Integration 27

4 The Drivers of Project Performance 35

5 The Drivers of Product Performance 75

6 Problem-Solving Foundations 99

7 Managing Technological Change 121

8 Building Technology Integration Capability 149

9 Integrating Technology and Market Streams 173

10 Rethinking R&D in a Chaotic World 209

Appendix I: Technical Analysis of Processor Modules 217

Appendix II: Regression Analysis 222

Bibliography 225

Index 235

About the Author 249

Preface

Our technological heritage is one of our most valuable assets. The results of decades of research in areas such as materials, electronics, computers, and software shape almost everything we do, from telephoning our parents to skiing down a mountain slope. And yet, research and development laboratories throughout the U.S. are under severe pressure to justify their existence as they go through restructuring, downsizing, or even as they are being shut down.

I have felt both the excitement and frustrations of research in a very real way. Before joining the Harvard Business School I obtained a Ph.D. in solid state physics, specializing in the fields of microelectronics and superconductivity. My own research focused on extremely small, superconducting electronic circuits. Although the work was fascinating and conceptually challenging, I was frustrated that its application to a real product often seemed too far off to even contemplate. Moreover, the world of research was changing around me. When I started my thesis in the early 1980s, research laboratories were vibrant institutions where scientists had plenty of money, worked on anything they wanted, and had virtually unlimited freedom in how they published their results. By the time I finished my postdoctoral work in 1989, most labs had a hiring freeze, their budgets had been curtailed, and their scientists were under strong pressure to find applications for their work.

I do not believe that these changes are the result of any decrease in the value of research. Research is, if anything, more valuable than it has ever been—IBM saves hundreds of millions of dollars each year in capital equipment expenditures just in its microelectronics business because of a number of breakthroughs achieved by its research scientists during the 1980s; Bell Labs innovations in speech recognition are worth even more to AT&T; postdoctoral research performed at Cambridge University laid the groundwork for the fastest search engines to be found on the Internet. But the world that current research must impact is more complex than ever, characterized by sophisticated competitors, fragmented markets, and billion-dollar production facilities. This means that translating the knowledge generated by research into a real, competitive product has become a very difficult task. Without an explicit process that reaches across a number of functions in the company and involves significant, dedicated resources, research results too often will never see the light of day. I believe that this is the reason why research laboratories are undergoing such turmoil. Not because research has no potential value, but because managers either do not recognize what it would take to make such a process work or are reluctant to take on the substantial organizational challenges this would involve.

This book is the product of eight years of my own research that has been aimed at understanding what organizations can do to manage the interaction between the worlds of research and product development. Its results are drawn from interviews and discussions with more than five hundred scientists, managers, and engineers in more than fifty computer and software companies. During the course of this work I found many technological opportunities whose enormous potential lay untapped. I found many traditional R&D processes that were slow and inefficient and hindered managers' attempts to match opportunity with market potential, or too distant from the contexts of emerging businesses for managers to recognize new possibilities. But I also observed several organizations that were outstanding at leveraging the potential of science and technology. The evidence I gathered shows that the emergence of new technologies can provide dramatic opportunities in dynamic environments. But to be used effectively, new technological possibilities must be carefully matched to their application context. Doing this well requires a proactive process that I call *Technology Integration.*

Performing this study would not have been possible without the help and encouragement of many managers, scientists, and engineers in the companies I studied. I am extremely grateful to all the study participants who spent long hours with me explaining the intricacies of their research and development processes. I would particularly like to thank: Don Seraphim, Rao Tummala, Trey Smith, and Bijan Davari from IBM; Gerry Parker, Craig Barrett, and Bob Jecman from Intel; Greg Blonder, Dan Krupka, Rich Howard, and Jim Clemens from Bell Laboratories; Masahiko Ogirima, Kanji Otsuka, and Kiichiro Mukai from Hitachi; Michiyuki Uenohara, Toshihiko Watari, Yuzo Shimada, and Mitsutoshi Itoh from NEC; Shun-Ichi Sano and Hisashi Hara from Toshiba; Rick Bahr, Brett Monello, Ed McCracken, Ron Bernal, BAM, Rajiv Deshmukh, Dan Lenosky, Rick Altmeyer, and Keith Mattasci (as well as all the engineers hanging out in the lab when Lego was being brought up) from SGI; John Zurawski and Jay Grady from DEC; David Yen and Ken Okin from Sun; Zack Rinat, Yarden Malka, and Nanda Kishore from NetDynamics; Tim Brady from Yahoo!; Julie Herendeen and Debbie Meredith from Netscape; and Roy Levien from Microsoft. I would also like to thank Mike Tinkham, my Ph.D. advisor, for teaching me everything I know about the process of scientific research.

This work would also not have been possible without the help of several outstanding doctoral students and research associates. Tarun Khanna, Jonathan West, and Alan MacCormack focused, respectively, on mainframes, semiconductors, and workstations, and made an infinite number of conceptual and practical contributions. Each of them was invaluable. Additionally, Warren Smith and Professor Paul Williams (the latter from Loughborough University, England) helped perform much of the field work. Paul Clark was very helpful in analyzing semiconductor data and assembling field evidence on Intel. Ellen Stein was brilliant in her ability to capture SGI's and Microsoft's culture. I would also like to thank the many members of the HBS Press who worked hard to ensure such a smooth and carefully executed publication process.

I am particularly grateful to a number of colleagues for mentoring and advice. Kim Clark and Steve Wheelwright were instrumental in encouraging me to focus on this issue and provided constant guidance and enthusiastic support. They helped frame the problem, shape concepts, communicate ideas, and generate company contacts. Eric vonHippel patiently worked with me to refine and expand many of my ideas. Gary Pisano, Takahiro Fujimoto, Oscar Hauptman, Marcie Tyre, Stefan Thomke, Rebecca Henderson, Anne Carter, Stefan Schrader, Dorothy Leonard, Kent Bowen, Bob Hayes, Clay Christensen, Mike Tushman, Mike Cusumano, Dick Walton, Ralph Katz, and Jim Utterback provided many invaluable suggestions, as did many participants in the Harvard University and Massachusetts Institute of Technology Technology Management workshops over the years. I am also grateful to Gary Pisano, Eric vonHippel, Dan Krupka, Jim Utterback, and Alan MacCormack for many comments that greatly improved the structure and content of this book. Finally, without Barbara Feinberg this book would never have been written. Her advice, insights, and editorial suggestions were consistently superb. They deeply influenced the structure of the book and the concepts behind it.

I'm most indebted to my family. My mother, Laura, provided

continued help and encouragement. My daughter, Julia, provided the spark, warmth, and excitement that made this last year so enjoyable. She also showed me how the most advanced organizational processes for experimentation and learning are vastly inferior to the capabilities of a one-year-old. Most of all, my wife, Susan Pope, provided the loving support that made this research possible. Her encouragement made it possible for me to manage the transition from physics to management. She was a constant source of ideas about concepts and methodologies, and her Ph.D. thesis on the impact of experience in situations of extreme technical uncertainty provided a critical foundation for my research. Her love and enthusiasm gave me the inspiration to finish this project.

Technology Integration

Introduction

chapter 1

TECHNOLOGY DOES NOT work in isolation. Not only can technology not be separated from the activities that surround it, *a* technology cannot be separated from other technologies. Technologies act in conjunction with one another: They only add value as integrated systems. But that integration does not happen by itself. Rather, people make choices about what should be integrated with what and for which purpose. Technology integration represents this process of choosing among technological possibilities, to solve a product problem. It thus defines the interaction between the world of research and the worlds of manufacturing and product application.

Imagine developing a new microprocessor. Its design is linked to a vast variety of technology choices. From among these, people must decide which algorithms should be used to optimize its performance, select a material for its package, define routines to test its reliability, and choose a lithography technique for patterning its transistors. Each of these technological decisions is a crucial element in the design of the product. To design a good microprocessor, each technology must be selected and refined so that it works seamlessly with all others and with the context in which the product is to be manufactured and used. Each technology must thus be integrated with all other technologies and with its context of application.

Technology integration has always been important, of course, but the world currently offers unprecedented technological possibilities. New advances in biology, chemistry, information technology, and materials science, for instance, have created a multitude of opportunities. Yet the potential of such discoveries often lies untapped as organizations struggle with their turbulent environments. This is because individual technologies rarely define products. To be deployed effectively, new technological opportunities must be carefully selected to fit within an increasingly complex and uncertain application context. Scientific research does not fulfill this role, since it is traditionally structured to explore—to assess the potential of narrowly defined technological possibilities. Nor does product development, which is usually structured to implement—to complete detailed designs and manufacturing processes. Rather, the process of making technology choices rests in integrating the activities that define the interaction between the traditional roles of research and of product development.

In a world in which the technological base is stable or the application context is simple, technology integration is relatively straightforward. Technology choices are comparatively clear and can be made within traditional research and development organizations. Many business environments no longer have a stable technology base or a simple application context, however; that is, their technological base is novel, changing rapidly and unpredictably and offering many possibilities for improving products and processes, and their application environment is complex, comprising a variety of standards, manufacturing processes, user needs, and competitive requirements. This combination of novelty and complexity creates an enormous challenge for organizations from manufacturers of computers to manufacturers of detergents. Selecting and refining technologies is very difficult when the options available are many and change rapidly and when the complexity of their context implies an array of subtle interactions between each decision. As such, recent increases in the diversity of technological possibilities and in the complexity of their contexts have greatly complicated the interaction between traditional research and development tasks, sharpening the need for a proactive process targeting technology integration. Consider the following examples.

Intel's latest production facility cost more than $3.5 billion, mostly for process equipment—a third of which had never been used before. As the goal was to improve the density of circuits with line widths squeezed below the wavelength of light, novel approaches to lithography, etching, and planarization were necessary. The requisite equipment was combined to create a manufacturing process comprising more than six hundred processing steps, each of which had to work in perfect coherence to achieve the extremely high production yields targeted by management. In the early phases of the project, no one could predict which fundamental technologies would function in the future plant environment, let alone produce reliably once combined to create the new process.

Microsoft faced an equally daunting task with its Windows95 operating system. The basic product requirement seemed simple enough: plug and play. That goal, however, implied its functioning with an almost unimaginable number of system—that is, peripheral—application software combinations; each of the new technologies included in Windows95 would have to work seamlessly in any of these situations. The result would have to include literally millions of instructions and a wide range of technological approaches. Microsoft, in other words, faced the same challenge that Intel did: how to start with a large number of technological possibilities with uncertain impact on a very complicated system and come up with a product that worked seamlessly, reliably, and coherently.

For both Intel and Microsoft, a proactive process of technology integration was central to their effectiveness. Creating novel technologies was not their only challenge; internal research organizations and external suppliers could provide many possibilities. And development itself was not the major challenge. In these well-oiled organizations, when a technological path is forged, implementation can follow swiftly via well-established managerial processes. The challenge, instead, was figuring out *what* to do; it was figuring out what to choose from the extensive palette of available technologies so that the future system solution would be relatively easy to develop and would work coherently. In other words, the challenge was to determine how technologies should be integrated.

Deciding how technologies should be integrated and creating the building blocks of a proactive process for doing so are the subjects of this book, which explores these phenomena in several computing-related industries. The work is founded on an extensive research base that includes four global empirical studies, each comprising field investigations of the major competitors in a focused industry segment. The research entailed field work on more than one hundred projects in semiconductors, mainframe and supercomputer subsystems, workstations and servers, and software.

Each of the industries studied represented different challenges —from the enormous capital investments needed for semiconductor production to the extreme market uncertainty of workstations and multimedia software. Nevertheless, each environment shared a need for an effective technology integration process. The capability of making good technology choices was a consistently critical contributor to R&D performance. Differences in the technology integration process could explain variations in project resources of a factor of three as well as delays of several years in the case of large mainframe projects. In semiconductors, the development of effective technology integration processes was associated with a major industry turnaround by U.S. competitors. In workstations and software, achieving a good match between technology and product architecture was critical to the products' competitiveness.

Further, differences attributable to a proactive technology integration process appeared to be at least as important as other managerial and organizational activities traditionally associated with studies of product development, for example, project management methods, leadership qualities, and the organizational structure of the development phase, among many others. The reason is simple: Once an organization has committed to a future product's concept, most of the potential for change and improvement is gone from the project. If the concept is a bad one, if the product is difficult to manufacture or inappropriate for the desired user application, the project will run into problems—no matter how well integrated the team or how powerful the project leader.

Technology integration is rooted in subtle organizational processes that merge the uncertain information on technological and market possibilities with the complex information describing the manufacturing and user environments. In a traditional R&D organization, technologies are chosen in a scattered fashion by individuals with varying experience, frequently before the project has gained any senior management visibility. By contrast, an effective technology integration process starts with an identified group of decision makers who have access to a rich, integrative perspective and are responsible for a broad, systemwide, technological outlook. Balancing individual requirements to define the best choice from a holistic perspective, they carefully craft a technological path. To inform the judgment of these people, the organization carefully nurtures their individual experience and emphasizes the need to gather knowledge of the user, product, and production system over several project generations. Critically, to assess the attractiveness of technological options before committing to a final approach, experimentation, prototyping, and simulation capabilities become central to a good technology integration process. The time for performing a full iteration between technology choice and system-level test is a critical determinant, as is the capacity for performing a large number of simultaneous experiments. These factors are linked to success even in turbulent environments like software and workstations.

Though the research concentrates on an industrial environment rightly perceived to involve complicated technological choices, the findings and implications are not limited to the computer industry. They should apply to any environment faced with a combination of technical uncertainty and environmental complexity, be it supercomputers or detergents. These findings imply a significant evolution in the structure of research and development in the modern corporation. The ability to leverage new science and technology effectively is related not only to the quality of the work performed in the scientific laboratory or to the ability to transfer and develop individual technologies, but also, critically, to *the capability of conceptualizing how a multitude of emerging possibilities might be used coherently to define a product that makes business sense.* This capability is at the heart of effective technology integration.

The book first analyzes the relationship among technology integration process, project, and product performance. It starts by setting up a conceptual framework for the analysis, in Chapter 2, and follows by describing, in Chapter 3, the structure of the empirical work. The book then dives into the empirical evidence, showing how the process of technology integration is associated with large differences in project performance (Chapter 4) and product outcome (Chapter 5). The analysis goes deeper in Chapter 6, exploring how project-level process is reflected in the workings of R&D organizations at the level of individual problem solving. This emphasizes the microscopic foundations of effective technology integration: matching knowledge of fundamental technologies with the complex characteristics of the application context.

The book then illustrates technology integration in a variety of environments and discusses the implications of our results. Chapter 7 describes several effective projects, some achieving evolutionary outcomes, others more revolutionary in nature, including detailed observations of projects at NEC and Silicon Graphics. Next, the development of technology integration capability is linked to characteristics of the firm's environment, such as the nature of the skilled labor market and access to research universities. The discussion is grounded in observations of the striking turnaround of the U.S. semiconductor industry, citing evidence from the progress made at Intel (Chapter 8). Chapter 9 illustrates the development of technology integration capability in the very different environment of software development. Rather than being large, traditional science-based firms accustomed to massive investments in manufacturing infrastructure, the firms looked at here are lean, young organizations, such as Netscape and NetDynamics, whose development projects require minimal capital investment. Still, the mechanisms driving technology integration capability appear to be similar. Indeed, this consistency itself is a major take-away from our research.

Science and advanced technology still lie at the heart of competition in a variety of environments. Evidence throughout this book shows, however, that an emphasis on creative research and focused development is not enough, as performance is consistently associated with the ability to manage their interaction through the integration of new technology. Chapter 10 thus concludes with a discussion of how the research recorded in this book challenges the traditional structure of the industrial laboratory, emphasizing the need for new approaches to the management of research and development.

The Concept of

chapter 2

Technology Integration

THE FACT THAT TECHNOLOGY is pervasive has become a cliché. Most of us take for granted that sophisticated, rapidly changing technology is the foundation for a vast variety of products and services we depend on, from integrated circuits in automobile brakes to financial models for managing pension funds. Nevertheless, although technology is everywhere, its development and application are still fraught with problems for both users and manufacturers.

While technical decisions are so common that they essentially run our lives, we still really do not know how to make them—and many organizations continue to make major mistakes. Consider a mainframe project from this study: It was delayed for two years because the packaging technology selected was, essentially, undevelopable. In another case, the architecture of a multimedia development project did not enable its on-line component to update its main application program. Both of these projects were such failures that their organizations left their respective businesses. Academic literature cites many examples showing, over and over, how organizations fail in the face of technological change (for example,

Abernathy and Utterback, 1978; Abernathy and Clark, 1985; Tushman and Anderson, 1986; Henderson and Clark, 1990; Anderson and Tushman, 1990; Christensen, 1992; and Christensen and Rosenbloom, 1995). Why is technological decision making such a challenge?

This book will argue that the challenge is rooted in a combination of novelty and complexity typical of rapidly changing technological environments. At a project's start, managers will not know for sure the precise nature of the options the new technology offers. Yet each of the options will affect multiple, interacting aspects of their future business. This interaction of novelty and complexity makes forecasting the impact of technology decisions exceedingly difficult—with frequent mistakes being the consequence. Moreover, the challenge must be addressed at a microscopic level. It is not enough to provide broad, strategic guidelines. Critical technology selection decisions are often buried deep within projects. Their impact is unrecognized. Problems come from unexpected sources, unforeseen interactions between hidden details. The two-year delay in the development of the mainframe mentioned above had a strategic impact *ex post.* Its causes, however, were lodged in materials choices made in the earliest project stages, before most of the senior management team was even aware that the project had started. The architectural problems in the software project had similar roots.

This book will also argue that technological decision making is an essential component of concept development activities in an R&D project and that the effectiveness of technical decisions is aided by a targeted process. That process should accumulate the appropriate knowledge base and allow it to be applied systematically. Evidence will show that applying the right resources, tools, and problem-solving approaches through the organizational process called *technology integration* provides enormous leverage for improving R&D performance.

Technical decision making thus complements external or customer integration, which Clark and Fujimoto (1991) argued establishes the critical link between customer and product concept (see Iansiti and Clark, 1994; and Fujimoto, Iansiti, and Clark, 1996). Understanding the customer is not enough. The creation of outstanding concepts requires selecting a feasible and effective technical path for translating objectives into reality. To achieve this, technological options should be carefully evaluated and selected in light of the complex requirements of their application context (see Figure 2-1). Clark and Fujimoto asserted that product concepts with integrity were not simply born from the minds of creative individuals, but

FIGURE 2-1

The Challenge of Technology Integration: Translating Technological Options into a Concept Matched to the Application Context

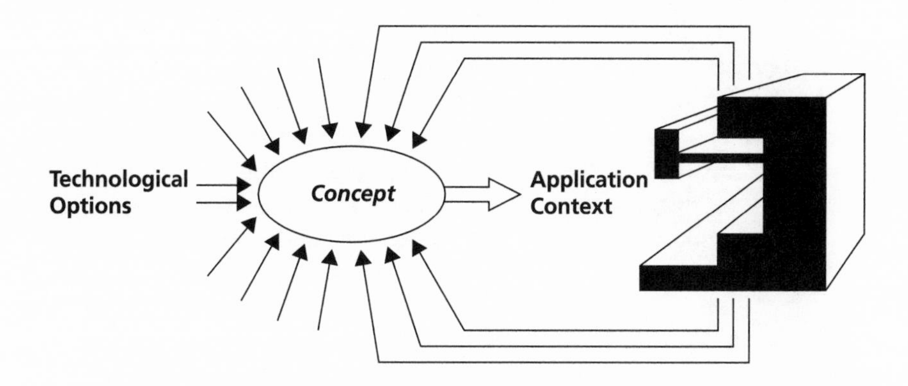

were the result of an effective organizational process. Similarly, technology choices that fit the requirements of a business cannot be made consistently by relying on the brilliance of only one or two scientists. Making good decisions requires knowledge, and the systematic creation, retention, and application of this knowledge requires a targeted process.

This chapter introduces the conceptual framework for technology integration. We first go to the roots of technological decision making and describe products as the outcome of knowledge-building activities. Traditional processes of research and of development are founded on certain basic premises about how knowledge ought to be accumulated and communicated, but these premises have been challenged by the speed and complexity of modern technological environments. The result is a gap between traditional research activities and development tasks, generating the need for the process of technology integration, which is defined in detail. The next chapter then sets the stage for subsequent empirical analysis by creating a framework linking elements of the technology integration process to the outcomes of the R&D activities—the performance of a project, the resolution of technical problems, as well as the nature of the product outcome.

Products and Knowledge

Products reflect the knowledge captured by the organizations that create them: This comprises knowledge of the underlying technical foundations, specific engineering fields, managerial processes, details of the manufacturing environment, users, channels, and markets. Our favorite armchair reflects knowledge of ergonomics; our personal computer reflects knowledge of circuit design and manufacturing. As such, R&D process effectiveness is linked to the evolution of the underlying knowledge base (see Thompson, 1967; Abernathy, 1978; Abernathy and Utterback, 1978; Rosenberg, 1982; Ettlie, Bridges, and O'Keefe, 1984; Tushman and Anderson, 1986; Henderson and Clark, 1990; and Von Hippel, 1990). Because this knowledge base comprises many different elements, we need to distinguish between knowledge of two different types: domain-specific, generalizable knowledge and context-specific, system knowledge.

Domain-Specific Knowledge

Part of the knowledge required for conceptualizing a new product is captured in domains. These are self-contained, fundamental disciplines independent of the immediate context of the product, such as the field of ergonomics or of transistor device physics. Domain-specific knowledge is relatively easy to articulate and generalize. The field of transistor device physics is built on analytical models, grounded in well-known and general properties of semiconductor materials. Ergonomics is founded on the study of human anatomy and is described in a variety of courses and textbooks.

Developing a product means drawing upon many domains. Knowledge needed for high-performance workstation design ranges from integrated circuit design to experimentation tools and techniques. Pharmaceutical development is based on advanced biochemistry and an understanding of the functioning of human organs, along with many other diverse specialized disciplines. Creating and retaining these diverse knowledge bases is a critical challenge. Effective product development is built on functional excellence, and no company in a competitive environment can survive without strong foundations in domain-specific expertise. Strength at the domain level is not enough, however. The diverse knowledge bases must be integrated with each other and with their application context to produce a product that functions consistently (see Marple, 1961; Alexander, 1964; Clark, 1985; and Von Hippel, 1990). In a workstation, for example, the integrated circuits must work with the circuit boards, and the whole must function reliably on the customer's software applications. This is where system knowledge comes in.

Context-Specific, System Knowledge

In outstanding products, diverse knowledge bases combine to create a coherent whole, well matched to its context of application (see Alexander, 1964; and Clark and Fujimoto, 1991). The knowledge needed to perform these integrative tasks is not usually captured by fundamental domains (see, for example, Rosenberg, 1982; Nelson and Winter, 1982; Clark, 1985; and Von Hippel, 1994). Part of the knowledge required to perform R&D is made up of integrative or system knowledge, which describes the interactions between the domains and their application context. System knowledge is by definition context-specific, and thus it is often difficult to articulate, or may even remain tacit (see, for example, Polanyi, 1958; Rosenberg, 1982; and Von Hippel, 1990). No textbook describes the details of particular manufacturing facilities. The manuals will capture only part of the information needed to design a new product. An employee with thirty years' experience in manufacturing will probably recognize that one design detail will be easier to manufacture than another, but she or he might not be able to articulate exactly why.

Collins (1982), for example, studied the development of the TEA laser. He described how several groups of scientists and engineers found it extremely difficult to replicate the published research of the original innovators. While the results of early experiments were published and widely disseminated, effective replication was consistently linked to the transfer of individuals with direct experience of laser operation from one project to another. Making lasers work depended not only on well-defined fundamental knowledge, but also on a wide variety of contextual details whose precise impact on the product was difficult to characterize and articulate. Von Hippel and Tyre (1995) studied this problem in depth in a manufacturing setting, discovering that new problems consistently appeared when a new piece of equipment was installed. These problems were caused by difficulties in transferring detailed knowledge of the manufacturing site upstream to the individuals charged with designing the equipment.

Research and development activities are thus aimed at the accumulation of specialized expertise in well-defined domains and the integration of these domains into a coherent system that works well in its application context. These objectives are founded on fundamentally different knowledge bases, and they require different approaches for their execution (see Burns and Stalker, 1961; and Lawrence and Lorsch, 1967).

Traditional Models of Research and Development

The existence of these two knowledge bases leads to contrasting problems in the management of innovation. First, knowledge domains change—frequently and in a rapid and erratic fashion. This creates the fundamental problem of *novelty*, which we define as the unpredictability of a knowledge domain (see Allen and Hauptman, 1987).[1] The more erratic its evolution, the less we know about what a knowledge domain encompasses. Combinatorial chemistry, for instance, is a rapidly changing discipline whose evolution is deeply influencing the options available to research and development in the pharmaceutical environment. While some experts feel that it will revolutionize pharmaceutical development, others think it will have only limited application.

The second fundamental problem in innovation management is complexity. Complexity can be defined as a high level of interdependence among domains (see, for example, Von Hippel, 1990; Allen and Hauptman, 1987; Tushman and Nadler, 1980; and Thompson, 1967).[2] The integrative challenges in designing a workstation, for example, are characterized by a high degree of complexity. The microprocessor interacts with the memory and with hundreds of other parts, including the application-specific integrated circuits (ASICs), circuit boards, and disk drive. Completing these complex projects involves coordinating the subsidiary activities to draw from a variety of interacting knowledge domains, such as ASIC design and circuit board simulation.

Traditionally, organizations have approached these sometimes contradictory challenges by partitioning the firm's innovation process into two distinct and separate phases. Novel tasks are performed in research, while complex tasks are performed in development (see Burns and Stalker, 1961; Lawrence and Lorsch, 1967; Moch and Morse, 1977; Allen, Tushman, and Lee, 1980; and Allen and Hauptman, 1987). Traditional R&D models work as follows: Research projects, aimed at the creation of technological possibilities, are optimized for the investigation of rapidly changing knowledge domains. Once enough is learned about these knowledge domains, research defines the technological possibilities available, which are transferred to the development organization. Development activities are optimized to execute complex tasks. These involve adapting the (now stable) set of technological possibilities to the complex requirements of the application context.

Research

Several studies have shown that research projects should have close contact with the changing base of disciplinary knowledge (see, for example, Allen, Tushman, and Lee, 1980; and Ettlie, Bridges, and O'Keefe, 1984), and benefit from deep specialization (see Allen and Hauptman, 1987). Research laboratories have thus typically been divided into narrowly specified functional groups. Each group focuses on building great depth in a rapidly changing domain of knowledge, such as the thermal properties of glass ceramics, the magnetic properties of two-dimensional superconducting systems, or the algorithms for conducting rapid searches in parallel computing (see, for example, Pugh, 1984; and Hounshell and Smith, 1988). This emphasis on deep specialization and narrow focus in research projects also reflects the scientific method. In order to approach highly novel problems, scientists are taught to minimize complexity by controlling for every possible external degree of freedom. The cleanest experiments are those designed to examine a narrowly defined, well-specified problem, such as measuring the speed of light in a vacuum or the magnetic moment of a superconducting material. The complexity of the outside world is therefore explicitly eliminated by isolating the experiment and providing controls.

Development

Development calls for completely different priorities. Rather than studying fundamental problems in isolation, the aim of development is to design something that functions robustly and reliably in the real world. Products need to fit with complex user needs and to be manufacturable by complicated production processes. Rather than building knowledge, development projects have, therefore, traditionally focused on execution. Rather than exploring novel domains, projects are optimized for managing complexity. Automobiles are a good example. The typical new car involves the redesign of millions of components, many of which interact with each other. The design of the door body panel influences the design of the interior trim and vice versa. The activities are complex, but not novel. Body panels and interior trims have been designed for many years, and the time scale of technological change is much longer than that of the typical project time line. Automobile product development projects have thus traditionally concentrated on implementation, not knowledge creation.

Academic research has shown that development projects are aided by close communication among individuals and coordination among tasks to resolve the many interdependencies. Therefore, in contrast to the high specialization and fragmentation of research, the use of project-focused organizational structures and processes has been linked to effectiveness in development (see, for example, Allen, 1977; Allen, Tushman, and Lee, 1980; Allen, Lee, and Tushman, 1980; Katz and Allen, 1985; Keller, 1986; Clark and Fujimoto, 1989; Fujimoto, 1989; Clark and Fujimoto, 1991; Cusumano, 1992; Cusumano and Nobeoka, 1992; and Bowen et al., 1993). Clark and Fujimoto (1991) investigated R&D processes in the automobile industry in depth, studying the impact of leadership, cross-functional interactions, and other factors. They showed these to be closely associated with project performance variables such as resources employed and lead time.

Technology Transfer

Traditional R&D models are completed by mechanisms for technology transfer. Several academics have studied the activities that provide a bridge between research and development. The research (for example, Cohen, Keller, and Streeter, 1979; and Leonard-Barton, 1988) shows that transferring knowledge from research to development (or from development to manufacturing) is facilitated by a variety of factors, including the transfer of individuals from research to the development organization, the existence of individuals in development who are broadly familiar with research outputs (gatekeepers, in the terminology of Allen, Tushman, and Lee, 1980, and Allen, Lee, and Tushman, 1980), and the adaptability of the downstream organization to the technologies developed in the upstream phases.

The technology transfer literature raises the issue of whether the upstream technology is well matched to the application environment. The work of Leonard-Barton (1988), in particular, describes the need for mutual adaptation between upstream researchers and downstream implementers. While there is a need for feedback between the two organizations, the emphasis remains on a linear process, with a clear partitioning of tasks. The work focuses on *a* technology that, if transferred effectively, will improve the performance of the downstream organization—not on what knowledge the downstream organization could build in order to choose which of many technologies would be appropriate in the first place. The upstream organization still focuses on knowledge building; the downstream organization still focuses on execution.

In sum, the literature has established a solid foundation for the study of research and development activities. It has described effective processes for research, for development, and for transferring knowledge from one to the other. It relies, however, on the premise that research and development activities can and should be partitioned cleanly, albeit linked by a technology transfer process—the essence of what we shall call the traditional approach to R&D (see Figure 2-2).

Inertia and the Emerging Gap Between Research and Development

The traditional approach to R&D does not always work. Many academics have documented the failure of R&D organizations to cope with significant technological change (for example, Burns and Stalker, 1961; Moch and Morse, 1977; Ettlie, Bridges, and O'Keefe, 1984; Abernathy and Clark, 1985; Clark, 1985; Dewar and Dutton, 1986; Tushman and Anderson, 1986; Henderson and Clark, 1990;

FIGURE 2-2

Two Models for Managing R&D

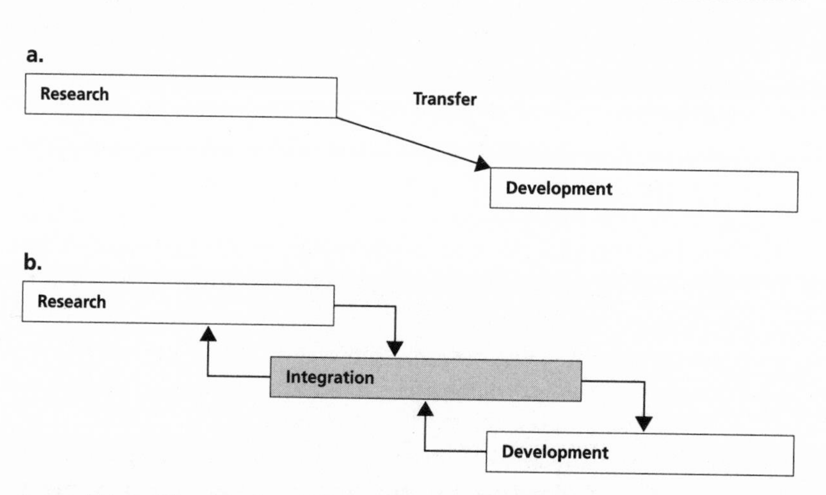

NOTE: a. This traditional model separates research and development but links them through technology transfer.

b. This alternate model, centered on technology integration, manages the interaction between research and development.

Anderson and Tushman, 1990; Utterback, 1994; and Christensen and Rosenbloom, 1995). Tushman and Anderson repeatedly found firms unable to cope with major technological shifts, in environments ranging from minicomputers to cement fabrication. Henderson and Clark showed that organizations failed to detect novel changes in the design of photolithographic equipment, even when changes were of a more subtle nature, that is, only partially shifting the knowledge base required for product design. Henderson (1994) later found similar issues in the pharmaceutical environment.

Hannan and Freeman (1984) argued that difficulties in responding to significant technological change were fundamental to organizations. These difficulties were linked to the routinization of tasks and to the associated phenomenon of organizational inertia. Routines would gradually become established through communication channels, information filters, and problem-solving strategies (see March and Simon, 1958; Arrow, 1962; Galbraith, 1973; and Daft and Weick, 1984).

Integrating the massively complex knowledge bases required for product and process development lends itself well to routinization. Development routines are thus created and improved through gradual refinements in times of technological stability (that is, in single-loop learning processes; see discussions in Argyris and Schon, 1978; Tushman and Romanelli, 1985; Levitt and March, 1988; Virany, Tushman, and Romanelli, 1992; Lant and Mezias, 1992; and Tushman and O'Reilly, 1997). Organizations increase efficiency by improving basic skills and standard operating procedures in efforts to optimize consistency and performance (see Clark and Fujimoto, 1991; and Wheelwright and Clark, 1992). Development routines thus become established and difficult to change, building considerable inertia.

This routinization of complex tasks in development leads to problems during times of significant technological change (see Tushman and Romanelli, 1985; Virany, Tushman, and Romanelli, 1992; and Levitt and March, 1988). And such change need not result in a total obsolescence of firm capabilities to cause serious problems. Henderson and Clark's work (1990) showed that the inertia built up by development organizations caused dramatic design failures even when only a small part of the technological base was affected by change.

The examples cited by Henderson and Clark are striking. In several cases, the organizations had even performed research in the specific domains undergoing evolution. They had studied the new fundamental technologies in the laboratory—moreover, they clearly had the capabilities to manage a project that would lead to the development of the new concepts. These organizations did not fail because of a fundamental inability to research and implement new technological approaches. They failed because they did not recognize the need for changing established development routines (see Henderson and Clark, 1990; and Christensen and Rosenbloom, 1995).

Following inertial trajectories that had solidified during past R&D activities, these organizations did not change *because they did not recognize the need to change.* Such recognition would have required the knowledge of what to do, as well as the energy and influence to make sure that the organization responded appropriately. These firms failed because both knowledge and influence fell into the gap between the traditional roles of research and of development. The research role focused on the evolution of individual knowledge domains, such as optics; the development role focused on getting products out the door through efficient (but rigid) routines. While researchers were working on the specifics of technological possibilities in a fragmented fashion, developers were too immersed in the existing context to recognize the potential of novel approaches.

Confronting the challenges associated with rapid technological change has become essential in a variety of environments. In Tushman and Anderson's study (1986) of the cement fabrication industry, times of technological upheaval punctuated thirty-year-long stretches of technological stability. In environments like that of semiconductors, major new technological possibilities come forward every five to ten years—virtually every major project generation. In Internet software, new possibilities come around on a monthly or even a weekly basis, that is, many times during the course of a single project. Rather than representing an occasional need, evaluating the impact of novel technological possibilities on a complex application environment has become an imperative for almost every major project.

This evaluation requires filling the gap between traditional research and development activities. In a complex and novel environment, making appropriate technology choices is difficult. Fighting inertia takes work. It means combining fundamental knowledge of all interacting domains with knowledge of the complex application context, applying the knowledge to challenge existing routines. Doing so in a consistent, sustainable fashion, project after project, is not a matter of having one or two smart or motivated individuals, but of having a *targeted process.* This is the objective of technology integration (see Figure 2-2).

Managing Technological Change

Inertial motion is not inevitable. In a physical setting objects will move in an inertial trajectory only in the absence of a counteracting force. In an organizational setting, inertial response is only inevitable in the absence of counteracting responses, such as learning (see March and March, 1977; Argyris and Schon, 1978; Nelson and Winter, 1982; Levitt and March, 1988; and Tushman and O'Reilly, 1997). *It takes work, but organizations do change.* Projects at Silicon Graphics, Microsoft, and Intel have repeatedly managed to get through technological transitions. While many of their competitors failed, these organizations adapted to different technological environments, customer preferences, and strategic priorities.

The evidence in this book shows that a process of technology integration can be instrumental in managing technological change and fighting the problems associated with organizational inertia. At companies like Silicon Graphics, Microsoft, and Intel, managing inertia is essential to competitive survival. These companies are thus masters at technology selection, and their experience is very useful in characterizing what makes technology integration work. Consider a recent example from Silicon Graphics (described in detail in Chapter 6): a high-performance server development project known in the company as *Lego.*

SGI's leading position in high-performance computing was being threatened by a variety of competitors offering less expensive products. The company's response was to design a revolutionary system with unprecedented performance, employing a completely new architecture—a similar approach had never before been implemented in the design of a commercial computer. Making the new architecture work involved integrating a variety of novel domains of knowledge (ranging from transmission line design to the thermal properties of integrated circuit packages) with a complex context characterized by a customized manufacturing process and highly sophisticated customer applications. Yet, the Lego team achieved its objectives. The product was shipped in the fall of 1996, gathering rave reviews. The entire effort took less than three years to complete. During this time, the new technologies were integrated into a seamless product that worked with a variety of sophisticated applications. How did they do it?

Several academics have argued that effectiveness during times of rapid technological change is linked to the skills and behavior of senior management. They showed that organizational learning and adaptation in times of technological change is influenced by a number of important factors, such as the composition of the managerial decision-making group, the resource allocation process, or the linkage between the firm and its market (see, for example, Leonard-Barton, 1988; Ancona, 1989; Virany, Tushman, and Romanelli, 1992; McGrath, MacMillan, and Tushman, 1992; Christensen and Rosenbloom, 1995; Tushman and O'Reilly, 1997; and Christensen, 1997). The senior management team can have a substantial impact on organizational response mechanisms. Executive succession and diversity in the executive team can facilitate second-loop learning and organizational adaptation (see Ancona, 1989; and Virany, Tushman, and Romanelli, 1992). Separating from the organization's mainstream any projects aimed at the development of disruptive technologies (Christensen, 1997) can increase their chances of success. Firms can thus both leverage established routines during times of stability *and* innovate through times of upheaval, what Tushman and O'Reilly (1997) call managing "ambidextrously." The success of Intel, Microsoft, and Silicon Graphics has been attributed in part to these factors.

Responding effectively to technological change cannot be the purview of senior managers alone, however. The relationship among new technology, product, and production system is defined in practice by a multitude of complex technology-selection decisions (see Leonard-Barton, 1992; Von Hippel, 1990; and Iansiti and Clark, 1994). Examples range from the selection of product materials to the choice of features; from decisions on design objectives to the choice of design tools and methodologies; from the selection of critical production processes to selection of advanced production equipment to be developed. The senior managers at Silicon Graphics could not, by themselves, oversee the many critical decisions that ensured that Lego really exploited its technological potential. These decisions may be microscopic in nature when compared to setting the strategic objectives of the firm, but when aggregated, they can have a critical impact on the performance and cost of future products, the speed and efficiency with which they are developed, and the overall competitiveness of the firm. Lego was not only successful because the company funded a project aimed at developing a novel computer architecture: Lego succeeded because its many technological approaches were chosen and refined so that they really fit together—and fit their application context. This meant that the technological approaches chosen could be implemented efficiently and on time, making possible a product with superior performance, exceeding the expectations of its customers. Lego's success, like similar successes on a variety of projects analyzed at other companies to be treated later in this book, reflected the technology integration capabilities of its organization.

Technology Integration

Critical technology-selection activities occur in any project that involves the implementation of novel technology. Technology selection is thus a bit like strategy setting. Organizations do it whether they know it or not (see Porter, 1980). In projects, critical technology commitments are made frequently, whether or not they are part of a targeted process, and whether or not senior management is aware of them. They happen whether they are the responsibility of a coherent group of individuals with appropriate experience, tools, and structure, or whether they are made in a scattered fashion by people in research, development, and supplier organizations.

The existence of a proactive, integrative process for making these choices is critical to R&D performance, however. The decisions represent a crucial window of opportunity for breaking inertia and managing technological change. Such microscopic decisions, for example, will define the cost of the new generations of DRAMs introduced by Texas Instruments; they will influence when Microsoft will ship the next generation of its Internet browser; and they will determine whether SGI's latest server will run efficiently as an Internet host. Mistakes may lead to failures as dramatic as industry exit. Decisions with such impact should not be made in a scattered and reactive fashion. They should be made in unison, in a proactive process that merges deep knowledge of the technological possibilities with detailed knowledge of the application context. The decisions should be made at an appropriate time and by an organization that has a systemic view of the process and the influence to make sure that any decisions made actually get implemented.

This does not mean that there is no role for research and development. On the contrary, the existence of a good process for technical decision making enables research and development activities to work more effectively, for both can be focused on what they do best. With research, this is the creation of individual technological options; with development, this is the execution of a well-defined product concept. Technology integration should thus not replace research or development, but *leverage both capabilities by managing their interaction.*

Technology integration is defined as the set of investigation, evaluation, and refinement activities aimed at creating a match between technological options and application context. The process should be integrative, since managing the relationship between new and old knowledge requires the fusion of a wide variety of information, from the fundamental science of computer architecture, for example, to the details of SGI's manufacturing environment. Technology integration should therefore provide the foundation for product development in a dynamic technical environment. The process should frame the project, providing a road map to guide design and development tasks. The activities should thus mediate between the intrinsically novel explorations performed in research and the complex processes of product and manufacturing process development.

Technology integration requires knowledge. This knowledge should provide the glue that holds a technologically diverse product together. Stated more formally, it should link fundamental knowledge domains to each other and to their context of application. This is knowledge of interactions—knowledge of how different disciplines influence each other: of how predictions by fundamental models of transistor performance are influenced by the actual characteristics of real production materials; of how advanced ceramic boards can be improved by coating them with polymers; of how software tricks can account for hardware problems in ASIC designs. In sum, the effectiveness of a technology integration process should be driven by knowledge that merges fundamental theories with the details of production systems and user environments. The elements of an effective process for technology integration thus fall into three types of mechanisms: mechanisms for knowledge generation, knowledge retention, and knowledge application.

Knowledge Generation

Knowledge generation will influence the quality of technology choices. The more we know before we commit to our choices about how technological possibilities will influence the context of application, the better the technologies chosen will work in practice. The greater the knowledge of the interaction between technologies and system, the better the match between the two.

Experimentation is the essence of knowledge generation. Many authors have emphasized the value of experimentation as a knowledge-building mechanism (see, for example, Arrow, 1962; Leonard-Barton, 1988; Adler and Clark, 1991; Rosenberg, 1982; Tyre and Hauptman, 1990; Von Hippel and Tyre, 1995; Pisano, 1996; Thomke, 1995 and 1996; and Thomke, Von Hippel, and Franke, 1997). Their work shows that experimentation is critical for learning and can play a central role in the evolution of organizational routines in situations characterized by rapid technological change. Much of this work has focused on learning curves and on the introduction of new process technology in a production environment. Adler and Clark (1991), for example, showed that learning through experimentation can lead to positive, systemic changes in the organizational processes involved in production. Several other academics (for example, Wheelwright and Clark, 1992; Bowen, et al., 1994; Thomke, 1996; and Thomke, Von Hippel, and Franke, 1997) have studied the impact of experimentation on R&D processes. Thomke showed that experimentation can substantially reduce development cost, for instance.

Although experimentation is always important in R&D, it is even more essential when the challenges are grounded in a combination of complexity and novelty. The work of Tyre and Von Hippel (see Tyre and Von Hippel, 1993; and Von Hippel and Tyre, 1995), for example, reveals that on-site experimentation plays an essential role in problem solving, given the complexity and "stickiness" of the information characterizing a production environment. Pisano (1996) also studied the role of experimentation in production process optimization, showing that the optimal experimentation strategy is influenced by the characteristics of the knowledge base describing the production environment. The less explicit the knowledge, the greater the role of experimentation in the actual production context. The same applies to other types of application contexts. Von Hippel has argued for many years that user environments are characterized by the challenge of sticky information.

This line of research therefore suggests that experimentation efforts that link technological possibilities with each other and with their application context should play a critical role in technology integration. Experimentation capability can be divided into three dimensions. The first is the *experimentation capacity*, defined as the number of different experiments an organization can perform in a given period of time. The second is the *experimental iteration time*, which is defined as the minimum time elapsed in a single experiment between its design and its execution. While experimentation capacity drives the number of parallel experiments an organization can run, experimental iteration time drives the number of sequential experimental cycles that can be carried out during a project. Both should be related to the ability to learn about the fit between technical options and context, and thus to project performance (see Thomke, Von Hippel, and Franke, 1997).

The third dimension of experimentation capability is the *representativeness* of the experimentation setup. This indicates the similarity between the experimental context and the actual context in which the product will function. The closer the similarity, the more indicative the results of the experiments, and the higher the performance of the project.

This dimension of experimental representativeness is a subtle one. An experimental setup will never perfectly resemble the actual context (see Von Hippel and Tyre, 1995). The complexity of a real manufacturing plant or user environment is usually too high to completely reproduce in a controlled setting. Moreover, the evolution of the product itself may influence its future context, motivating the user to develop new applications, for example. Designing an experimentation facility should, therefore, not only be driven by what the previous context looked like, but by aggressive estimates of the future context. A good experimentation facility for technology integration should allow for a diverse set of experiments, testing a broad variety of technology—context combinations to explore the potential for unexpected interactions.

Knowledge Retention

Effective technology integration cannot be based solely on knowledge generation, however. While experimentation can be instrumental to knowledge building, it is difficult to generate an exhaustive knowledge base in every project. Moreover, experiments need to be steered and their results interpreted. Managing technology integration, therefore, requires skills and capabilities through which knowledge generated via experimentation can be complemented by knowledge captured by individuals over time (see, for example, Cohen and Levinthal, 1990). Building this capability base generates the need to link different projects together, effectively orchestrating the evolution of the R&D organization's knowledge base.

An effective technology integration process depends on a solid foundation of system knowledge. This is not easily stored or transferred from individual to individual, however, as argued above (see for example, Leonard-Barton, 1988; Clark and Fujimoto, 1991; Von Hippel, 1994; and Von Hippel and Tyre, 1995). The value of individual experience in such situations is therefore very high (see, for example, Leonard-Barton, 1988). Effective organizations should retain people with the experience base necessary to investigate the broad impact of individual design decisions. These professionals should drive project specification, becoming the architects or integrators of the future product. Effective organizations will emphasize the growth of such experience over time and will value these individuals highly.

Several academics have also suggested that too much experience and too much rigidity in team structure and career paths can be problematic (see, for example, Katz, 1982; and Leonard-Barton, 1992). We can therefore expect that the impact on performance of such knowledge-retention mechanisms as experience may be subtle. Although some experience is essential to capture the nature of the context of application, too much experience might lead to inertia. Innovation often stems from diversity in the experience and skill mix of decision makers. New skills and perspectives can help provide the new frames of reference needed to examine novel technical possibilities and to question existing solutions. Thus, while some experience may be linked to effectiveness, too much experience might lead to rigidity, restricting the breadth of technical options considered in the decision-making process.

Knowledge Application

To be effective, knowledge retained through experience and generated through individual experiments must be integrated and applied. Hence, a project must have an organizational process that allows knowledge-generation activities to be coordinated and used coherently to make technology choices. This way, the consequences of individual decisions can be balanced against each other to make sure that the systemic impact of individual choices is assessed.

In practice, many organizations do not manage technology integration decisions coherently. The process is instead performed by a variety of engineering groups in a scattered fashion (see Hayes, Wheelwright, and Clark, 1988; and Wheelwright and Clark, 1992). New technical possibilities are then selected for their individual potential—not by emphasizing the formulation of an integrated system-level concept. In contrast, project performance should be associated with a focus on a proactive analysis of the systemic impact of technical possibilities, performed by a dedicated group of individuals.

Finally, the timing of these activities is important. Technology integration tasks provide a critical window of opportunity (see Tyre and Orlikowski, 1994) for managing technological evolution. Once the window is closed, the technological path has been selected and the project will rapidly build up momentum in the specified direction. At this point, inertia has already been created, and the options left open to managers and engineers will become severely restricted. If the window closes too early, problems will arise as inconsistencies emerge between new technology and existing system characteristics. These will cause delays and require the commitment of additional resources, leading to poor R&D productivity and longer lead times.

In summary, a good technology integration process should proactively induce a broad and informed approach to decision making and problem solving. The process should emphasize experimentation aimed at the early generation of knowledge about the potential impact of novel approaches on the application context. It should manage the retention of knowledge through past experience. The knowledge retained through experience and generated by experimentation should be integrated by a dedicated group of individuals charged with making technology choices with influence over the relevant application context. The choices should be kept open until informed decisions can be made, thereby avoiding premature commitments.

The next chapter focuses on the structure of the empirical work, showing how the research was designed to investigate these claims.

Notes

1. Allen and Hauptman actually define novelty as *the rate of change* of knowledge. My definition is therefore slightly different in its reference to the ability to predict or forecast the evolution of a knowledge domain at a point in time. This appears to better reflect the nature of novel technological change, which is often perceived as discontinuous.
2. Others have frequently defined complexity as interdependence between *tasks* instead of domains. This definition would not change the argument appreciably, since one almost inevitably implies the other.

chapter 3

Studying Technology Integration

ANALYZING THE RELATIONSHIP between technology integration process and performance in a novel and complex environment entails difficult challenges. How does one develop accurate and robust measures of the performance of each project? How can one guarantee that the sample of organizations includes enough variation in approach to be interesting and representative? How can one define a methodology that is structured enough to provide reliable quantitative analysis while being rich and flexible enough to capture the subtlety of the phenomenon one is trying to study? The very premise that the technological base evolves unpredictably and that the context of application is complex makes these questions difficult to tackle.

I approached these challenges by founding the research on a series of cross-sectional comparisons of projects, each limited to a very narrowly defined technological environment. Each environment was chosen to make sure that it had well-defined, stable, and reliable performance measures and that it included as broad a range of organizations as possible. I also used multiple empirical lenses to observe both performance and organizational process, combining extensive qualitative field studies with structured empirical comparisons. This structure created a robust and rich analysis, examining several empirical environments through multiple empirical methodologies. The goal was to show that technology integration is *consistently* linked with both project and product performance, as well as to shed light on the details of the processes that make technical decision making effective.

Field Work Design

Field work in each empirical environment searched for consistency by performing several independent analyses (see Figure 3-1). After first measuring the performance of each project, I analyzed the process that each followed and examined the relationship between technology integration process and project performance. I then characterized the outcome of the project to see if "integrated" processes indeed were reflected in "integrated" products. Finally, zeroing in on the microscopic problem-solving behavior in each organization allowed testing of whether integrated processes at the project level were in fact reflected in integrative daily problem-solving behavior.

Project-Level Process and Project Performance

The core of this empirical work examines the relationship between the processes employed by an organization at the project level and the performance of that project. This methodological approach is rooted in the work of many others (see, for example, Allen, 1977; Von Hippel and Tyre, 1995; and, especially, Clark and Fujimoto, 1991). I first developed a set of project performance measures, such as project resources or lead time, which were representative of factors critical to firm performance in its competitive environment and then developed indicators of organizational process. I then tested the hypotheses developed in the conceptual work by correlating process with performance outcome.

The study of technology integration challenged this methodology in several ways. First, given the novelty in the technological base of the environments studied, comparing the performance of different projects in a precise way was a formidable undertaking. Comparing the development of a personal computer to the development of a mainframe, say, could not be done by simply adjusting for the price of the product or the number of parts, since each product comes from a fundamentally different technical base. While the

FIGURE 3-1

Empirical Design: Searching for Consistency Across Levels of Analysis and Performance Measures

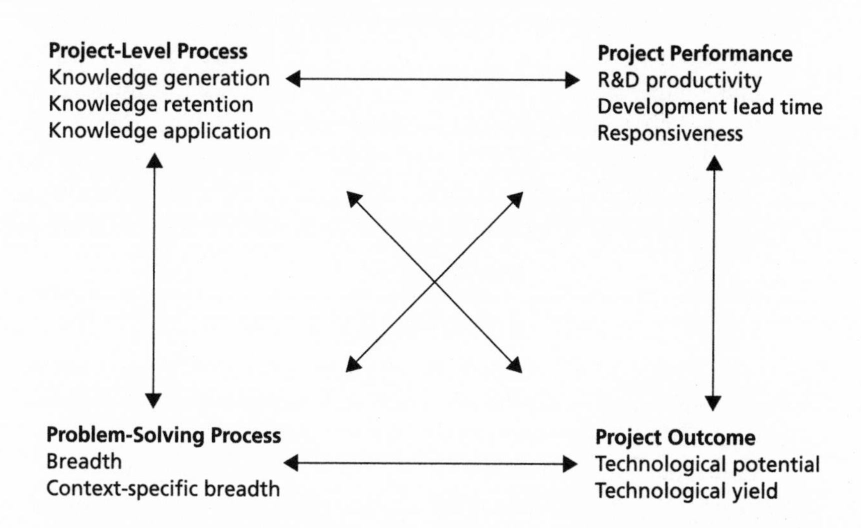

development of a new mainframe hinges on novel technologies, the development of a personal computer makes use of existing components. Since the goal was to study the relationship between organizational process and performance, it was necessary that basic differences in the technological environment of a given project not drive any observed difference in project-level process and performance outcome.

To solve this problem, I decided to perform the field work in a series of very narrowly defined industry segments. This ensured that each project comparison was truly made in a level, apples-to-apples fashion. For example, the study of mainframe development focused only on one critical component: the processor module. Each of the projects studied faced comparable technical novelty and complexity. What's more, because of the nature of the processor package, each project's output could be compared precisely along well-defined performance dimensions. A similar approach was used in the studies of semiconductor components and workstations, so that each time a comparison was made, observed differences in performance could confidently be linked to differences in organizational process—not to basic differences in the environment surrounding each project.

Next, I needed to find ways to assess differences in the process of technology integration. The full-scale empirical research was thus preceded by a year-long pilot study (Iansiti, 1991). During this time, I investigated two organizations in detail, one based in the United States and one in Japan, both involved in the development of mainframe processor modules. These investigations were used to develop the basic empirical frameworks for the study, including the definition of precise indicators describing the mechanisms for knowledge generation, retention, and application outlined in Chapter 2. These indicators are discussed in detail in Chapter 4.

After the end of the pilot phase, I collected observations on the major projects recently performed by firms in the mainframe computer industry. The sample comprised *all* companies in the industry, including U.S., European, and Japanese firms. In each company, at least one project was analyzed in detail, for a total of twenty-seven projects. Structured and unstructured interviews were held with scientists, engineers, and managers at different levels in the organization involved with the most critical aspects of each project. A questionnaire was used to guide the interviews and to gather additional data to add to and check the information drawn from the interviews. Histories were recorded for each development effort, tracking the completion of each major step as well as the resources used, and observations were gathered on the basic characteristics of the organizations, the processes employed, and the behavior patterns of the managers and engineers. These observations were used to create estimates of R&D performance and project-specific indicators of technology integration processes.

A similar methodology was repeated for the semiconductor and workstation environments. In total, detailed field observations were gathered for more than one hundred projects. Chapter 4 describes the results of this approach in relating project-level process and performance. As anticipated, the results show a clear association between project performance and the mechanisms of knowledge generation, retention, and application.

Project Outcome

Technology integration is aimed at obtaining a "good" match between technology and application context. But how do we assess the goodness of the match? How well did the technologies chosen actually fit with the application context? Answering this question meant developing a methodology for evaluating the quality of the final match between technology and context. To do this meant returning to the original notion of a product as the integrated outcome of both fundamental and system knowledge. Creating a product, as argued in Chapter 2, hinges on fulfilling two critical objectives: acquiring a foundation of fundamental knowledge and integrating this knowledge into a complex object that functions as a coherent system in a real context. The methodology I developed and demonstrate here thus assesses technology choices by measuring how well a project achieves both of these objectives.

This methodology divides product performance into *technological potential* and *technological yield.* Technological potential is an estimate of the product's maximum potential performance given its base of fundamental knowledge. It is drawn from an analytical model based on the physical laws describing the product. It answers the following question: Given the fundamental technology choices made, what is the maximum possible performance that the project could achieve?

Technological yield measures the extent to which the product's potential is translated into actual performance. This probes how well a given base of fundamental technology is integrated with the context of application, represented by overall product architecture, production, and user environment. In other words, technological yield measures how close to the maximum theoretical performance the project actually comes.

These two variables provide much information on the nature of the technical approach a project takes. A project with high potential and low yield is characterized by aggressive choices at the fundamental level. It will, however, have missed opportunities in the integration of these technologies into a real product. A project with low potential and high yield is characterized by choices that are more conservative at the fundamental level but better leveraged into product performance.

Technological potential and yield, therefore, assess the outcome of technology integration activities. While the potential measures how far fundamental technologies have been pushed, the yield estimates the closeness of the match between fundamental technologies and application context. Chapter 5 applies this methodology to the empirical data and shows that development projects obtain significantly different results in technological potential and technological yield. As expected, technological potential and yield are linked to technology integration process characteristics, such as mechanisms for knowledge generation and retention.

Problem-Solving Process

Problem-solving activities are an engine of technological evolution (see, for example, Dosi and Marengo, 1993; and Iansiti and Clark, 1994). They drive the evaluation of new technological possibilities and the generation of new knowledge. The problem-solving behavior observed in a project is therefore a wonderful lens through which to observe the process for making technology choices.

Problem-solving processes have been studied extensively by a number of authors (see, for example, Simon, 1978; Frischmuth and Allen, 1969; and Mintzberg, Raisingani, and Theoret, 1976). Frischmuth and Allen developed a model of technical problem solving emphasizing two streams of activities: the generation of solutions and the generation of criteria (or frames) for evaluating those solutions. They pointed out that a characterization of both activities is essential in understanding engineering problem-solving strategies. Using broad, integrative frames of reference is particularly important in a novel and complex environment (see Tushman and Romanelli, 1985; Dosi and Marengo, 1993; Schrader, Riggs, and Smith, 1992; and McDonough and Barczak, 1992), because the combination of novelty and complexity creates ambiguity. When we start evaluating new technology, not only do we not know the answers, we frequently do not know exactly what we must do to discover them.

The methodology for analyzing the disciplinary breadth of problem solving is described in Chapter 6. It entailed capturing detailed case histories of more than sixty critical problems encountered and analyzing the number of separate knowledge domains that each process went through. The more the domains, the broader and more integrative the effort.

The analysis shows that problem-solving behavior correlates with project-level behavior and performance. Integrative behavior in problem solving is indeed mirrored in the match between technology and context and in the macroscopic processes used at the project level. This confirms the relationships postulated in Figure 3-1.

Empirical Settings

The analysis of project characteristics, product outcome, and problem-solving behavior is conducted in a variety of empirical settings. The empirical work focuses primarily on the four environments shown in Figure 3-2. Each is characterized by the simultaneous challenges of technical novelty and environmental complexity. All projects faced rapidly changing and ambiguous technological possibilities as well as a diverse and complicated context in which the technologies needed to be applied.

FIGURE 3-2

Empirical Settings

Personal computer and Internet software	*Software*
Workstations and servers	*System*
Mainframe processor modules	*Subsystem*
Semiconductor components (microprocessors, DRAMs)	*Component*

The nature of novelty and complexity in each of the environments varied considerably. As Figure 3-3 indicates, in going from component to subsystem and from system to software, the complexity of the manufacturing environment decreased and the complexity of the user environment increased. In semiconductor component manufacturing, the manufacturing process includes hundreds of steps and equipment costing billions of dollars in a state-of-the-art facility. In software, the manufacturing process is almost trivial. Conversely, DRAM customer requirements are simple and stable, evolving along a predictable trend line. In workstation development, on the other hand, customer requirements are subtle, fickle, and diverse. Moreover, the nature of the technical challenges and disciplines involved also changed drastically from environment to environment. In semiconductors, the critical disciplines were solid state physics, chemistry, and materials science; in workstations, ASIC design and electrical engineering; in software, mathematics and computer engineering.

It must be emphasized, therefore, that although each environment is associated with the computer industry, the differences among empirical settings are very substantial. As such, comparisons between DRAM and mainframe subsystem development and between workstation and software development are useful tests of the generalizability of the results obtained.

A Consistent Outlook

In summary, this book presents a broad base of empirical evidence and discusses observations made by a variety of independent

FIGURE 3-3

Sources of Complexity in the Empirical Environments

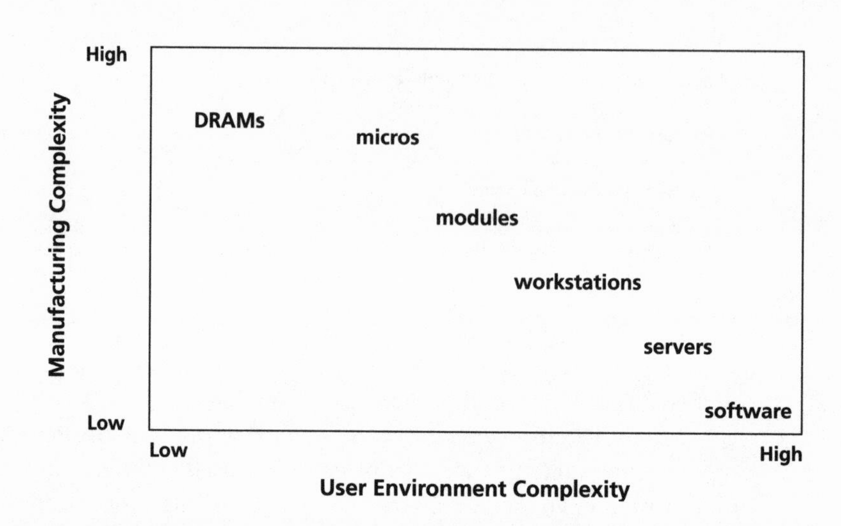

methodologies assessing project characteristics, project and product outcome, and problem-solving behavior. The evidence covers settings as different as a multi-billion-dollar semiconductor process development project and an Internet software development project carried out by a handful of engineers over a few months. The findings present a consistent outlook, across industrial environments, methodologies, and levels of analysis. The outlook suggests that technical decisions should not be approached in a scattered and reactive fashion. Rather, technology integration is a central challenge in a novel and complex world, and the process for matching technologies to the application context is associated with dramatic differences in both project and product performance.

The Drivers of Project Performance

chapter 4

WHEN I BEGAN the empirical work covered here, I expected to find that different competitors in the same environment would achieve comparable project performance and exhibit similar organizational processes. Although I expected to find *some* differences (as in Clark and Fujimoto, 1991), I did not think these would be large, especially where all of the projects were developing similar products and facing similar technical problems. Instead, I found differences in performance and process among projects were vast and critical to firm competitiveness. Consider two mainframe processors, designed by different firms: the same final product performance, similar architecture, same reliability, same standard interfaces, same customer base. Yet one took four years and fewer than two hundred person-years of effort, while the other took six years and more than *five* times the person-year resources. The two projects also employed completely different organizational approaches to technology integration, valued different experience bases, and used different experimentation methodologies.

The same types of results were common to the entire sample. When the empirical work was planned, the first concern was to ensure accurate, apples-to-apples comparisons among projects and organizations. Nevertheless, no matter how narrowly the empirical work was focused, dramatic differences were discovered among competitive firms. There were differences in the lead time of the projects and in the productivity of the organizations. Moreover, competitors approached technology integration in fundamentally different ways, emphasizing different tasks, skill sets, organizational processes, and types of tools and infrastructure. Even when two organizations tried to solve exactly the same technical problem, they frequently followed very different paths to do so.

Throughout this book, I will argue that such variation is driven by fundamental differences in how organizations approach the integration and implementation of new technology. Some projects emphasize the pieces, optimizing the impact of individual technologies, thereby making selection decisions in a fragmented fashion. Others view the technology-selection problem in an integrated fashion, maximizing the aggregate impact of all choices on the entire system. And though it requires substantial investment in infrastructure and capabilities, this second philosophy is associated with superior performance in rapidly changing technological environments.

This chapter launches the journey through empirical observation, examining the relationship between process and performance at the project level and introducing data from several studies performed in different empirical environments. Although each project was characterized by individual challenges, the process for technology integration was consistently associated with the large differences in project performance. Before setting out, however, it will be useful to recall several points made earlier.

Chapter 2 argued that effective projects in a novel and complex environment should employ a targeted process aimed at the generation and retention of system knowledge and at its application to technology-selection decisions. Three important types of project-level mechanisms for creating a good match among technology, product, and production system were cited:

1. Mechanisms for the *generation* of knowledge of technology-system interactions before commitment to a given technological path (especially through the availability of tools and processes for experimentation).
2. Mechanisms for the *retention* of knowledge of the product and production system over time (especially through the buildup of individual project experience).

3. Mechanisms for the systemic and timely *application* of the knowledge generated and retained (through the existence of a group of individuals dedicated to this task).

This chapter investigates the association between these mechanisms and project performance, analyzing project-level evidence from the various empirical studies conducted. Overall, the findings provide a compelling argument for the importance of technology integration as a fundamental process in novel and complex environments.

Process and Performance in Mainframes

The first evidence comes from the world of high-performance computing. The research, conducted between 1989 and 1992, focused on the development of the last major, all-out generations of high-end mainframes. At the time of the study, the mainframe computer industry represented about $30 to $50 billion in annual sales and was intensely profitable and competitive.

Empirical Environment

Mainframe computers are a good environment for studying technology integration; they are highly complex, and product performance has grown rapidly, largely through the introduction of novel technologies. To obtain precisely comparable observations, I focused on technologies related to the multichip module, which contains the mainframe processor. This subsystem is critical to the computer's speed and reliability. In the words of a senior executive at one of the leading companies in the area, this is "the most critical subsystem in the product, and among the most difficult to develop."[1]

New modules frequently draw on emerging scientific concepts such as novel ceramic materials. They are also very complex, requiring a wide variety of knowledge bases in their development—from material science to heat diffusion modeling. And they are composed of a large number of interacting components, some aimed at cooling the integrated circuits, some at protecting them, and others at transmitting their electrical signals (see Tummala and Rymaszewski, 1988).

Project Performance

The results of the study of project performance in mainframe processor modules are summarized in Table 4-1 and in Figure 4-1. Showing data on lead times, person-years, and project content, the

TABLE 4-1

Average Measures of Project Performance and Content in Mainframe Module Development (Standard Deviations in Parentheses)

	Total	Japan	U.S. and Europe
Number of projects	27	11	16
Project performance:			
Total lead time (years)	8.1	7.6	8.4
	(2.8)	(2.3)	(3.1)
Concept lead time (years)	3.2	3.7	2.9
	(1.6)	(1.5)	(1.6)
Development lead time (years)	4.8	3.9	5.3
	(1.9)	(1.2)	(2.1)
Project resources (person-years)	347	188	456
	(386)	(104)	(469)
Project content:			
Technical content	0.39	0.44	0.36
	(0.33)	(0.35)	(0.33)

Differences in development lead time and person-years between Japan and the United States and Europe are statistically significant at the 5 percent level.

table is organized by geographic origin to indicate the variation while disguising the performance of individual organizations. (U.S. and European companies are grouped together to disguise the characteristics of individual organizations, since there were only three European companies in the sample.) The data set comprised every major organization involved in this business around the world and all major projects concluded between 1980 and 1990. The figure summarizes the person-year data.

The results are striking. Project resources differ by almost a factor of three between the two geographical groups. Differences in development lead time represent years.[2] Moreover, the spread within each group is also quite large: There was a variation in efficiency between the most and least productive projects that approached a factor of ten.

What do these measures mean? The most telling project performance measure is person-years. This measure (when adjusted for content) focuses on the resources needed by an organization to research and develop a typical product. The smaller the number, the

FIGURE 4-1

Differences in Human Resources Needed to Complete Mainframe Processor Module Projects in the U.S. and Japan

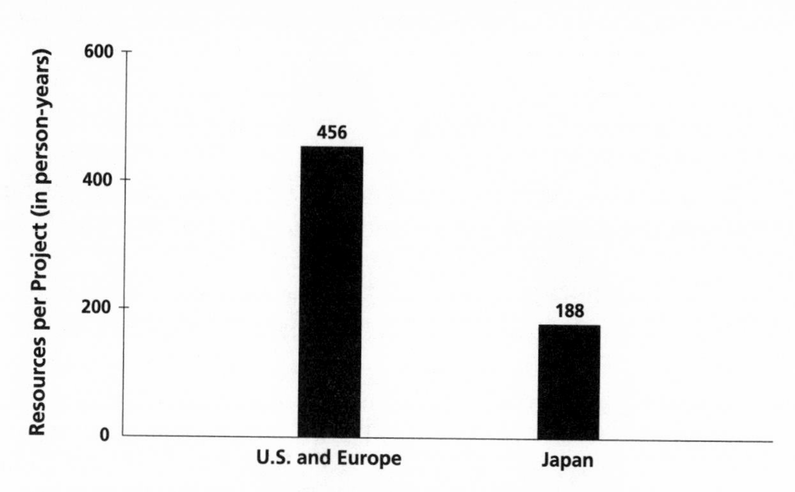

NOTE: Averages of resources per project used in a sample of 27 projects.

higher its productivity, and (given fixed, or sticky human resources) the shorter the cycle time between product generations. The fewer the content-adjusted person-years, therefore, the higher the organization's rate of technical progress for a given resource level.

The differences in project resources shown are critical to the competitiveness of the firms. The resources required in this type of activity are exceedingly scarce, difficult to find externally, and difficult to develop internally. Availability of highly skilled employees is typically the bottleneck in each project, constraining the number of products that can be developed on a yearly basis. High productivity means being able to introduce more innovative products in a given period of time.

Lead-time variables are also important. The measure most representative of speed is the development lead time: the elapsed time between commitment to a technological concept and its market introduction. It therefore indicates the minimum time between the integration of a new technology and its market launch. The shorter the development lead time, the easier it will be to include the latest major technical (or market) innovation in the current product generation.

Concept lead time instead indicates the length of time between the earliest technical explorations and the end of the technology integration phase, effectively measuring the size of the window of opportunity for integrating novel technologies. A long concept lead time can therefore be a positive project characteristic, since the wider the window, the greater the opportunity for including novel technologies in a project. Total lead time is composed of the sum of concept lead time and development lead time; a long total lead time can therefore be either good or bad, depending on the ratio between concept and development lead time. I will therefore focus on project resources and development lead time as the central measures of project performance. The smaller the resources and the shorter the development lead time, the better the project.

The differences among the projects in the table cannot be explained by variation in project content (that is, the fact that easy projects take fewer resources than do difficult projects). Given the focused way in which the empirical environment was selected, content differences are small. Moreover, compensating for them actually goes the other way. Adjusting for the small differences in content *increases* the differences in resources and lead time between geographical groups, as shown in Figure 4-2. The variation is also not explainable by supplier contributions (that is, by suppliers

FIGURE 4-2

The Impact of Project Content Adjustments on Person-Year Averages

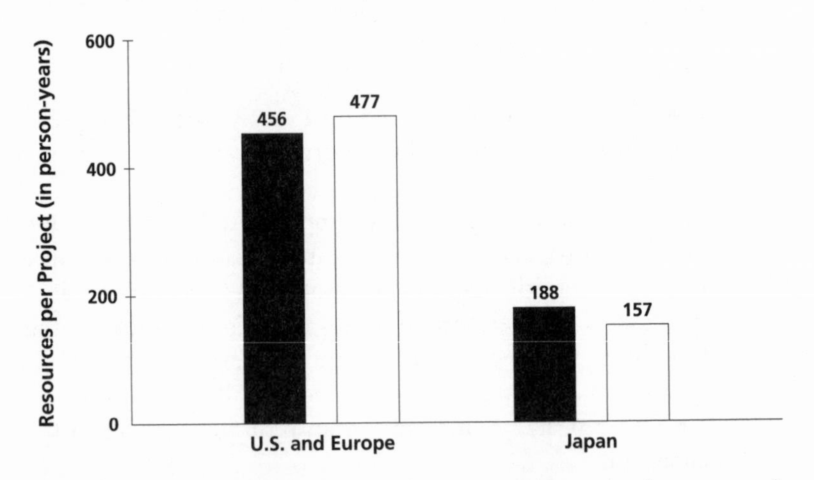

NOTE: The black bars indicate the averages of resources used per project (in person-years). White bars indicate the impact of content adjustment on person-year averages.

TABLE 4-2

Summary of Performance Variable Definitions

Total lead time	Time elapsed between the beginning of the project and market introduction. The project beginning is defined as the start of the first scientific investigation of new technologies specifically targeted for possible inclusion in a new design.
Concept lead time	Time elapsed between the beginning of the project and the end of the technology integration phase. The latter is signaled by the emergence of a firm and detailed technological concept for a new system.
Development lead time	Difference between *total lead time* and *concept lead time.* Indicates the time spent in development. During this phase, activities are focused on a detailed design of the product and associated process, based on specifications established during the integration phase.
Project resources	Level of technical and managerial human resources used over the entire length of the project. Includes engineers, managers, scientists, and technicians—both internal and external to the firm.
Technical content	Ratio of the number of gates per square centimeter achieved by the project to the best value achieved in the industry (best in class) at the time of product introduction. The measure is a good indicator of the overall functional performance of the device, since it depends on the most critical characteristics of the design, such as electrical, thermal, and mechanical characteristics.

doing part of the work), since these are included in the person-year measures.

Plotting the performance of individual projects reveals even larger differences in development lead time and resources than those between geographically determined averages. This is illustrated by plotting the residuals of regressions of technical content with person-years and development lead time (see Figure 4-3).[3] The numbers show how much variation is left over after differences in project content are included. The amounts are staggering—hundreds of person-years and several years in elapsed time differentiate the best from the worst. These differentials are wider than those found in studies of more mature environments, such as the automobile industry (see, for example, Clark and Fujimoto, 1991).

Figure 4-3 shows an additional startling fact. It indicates that the speed of development is *inversely* proportional to the resources allocated—the more person-years used, the longer the development lead time.[4] This result contradicts the traditional assumption that the more resources assigned to a project, the faster it proceeds. Instead, the opposite happens. Some projects have short development lead times and use relatively few person-years; other projects have high values of both. Rather than a simple tradeoff between time and resources, there are large differences in the overall effectiveness of projects.[5] In other words, some projects work well and complete the development of new products quickly and efficiently. Others do not work well, but run into very significant delays and resource overruns. To understand the reasons, we need to look at the development process followed by different projects.

Relating Process to Performance

I next matched the approach followed in the projects with the results they achieved. This required several years spent in making multiple visits to the companies in the study in order to make detailed, in-depth observations.[6] Literally, one week would take me to IBM for interviewing and the next one would take me to Japan to look at Hitachi.

The first phase of the analysis looked at the impact of traditional drivers of development performance. Several authors have speculated about what drives project performance (see, for example, Allen, 1977; and Clark and Fujimoto, 1991), zeroing in on the impact of development. These authors have identified various factors that contribute to project success—communication among participants, cross-functional integration, project structure, and the skill and influence of the project leader. Their work showed that these factors are critical to performance in several environments. In contrast to many traditional settings, however, the empirical environments discussed here were characterized by much more technical novelty (as in the work of Henderson, 1994, and Pisano, 1996, on pharmaceutical development). As detailed in Chapter 2, technology integration was expected to play a more important role. Showing this, however, meant first analyzing the extent to which the more traditional factors might explain the observed performance differences between projects.

Table 4-3 lists some of the more traditional organizational variables included in this analysis. Following the work of Clark and Fujimoto (1991), I defined three indexes to summarize the general

FIGURE 4-3

Project Resource Residuals Plotted Against Development Lead Time Residuals

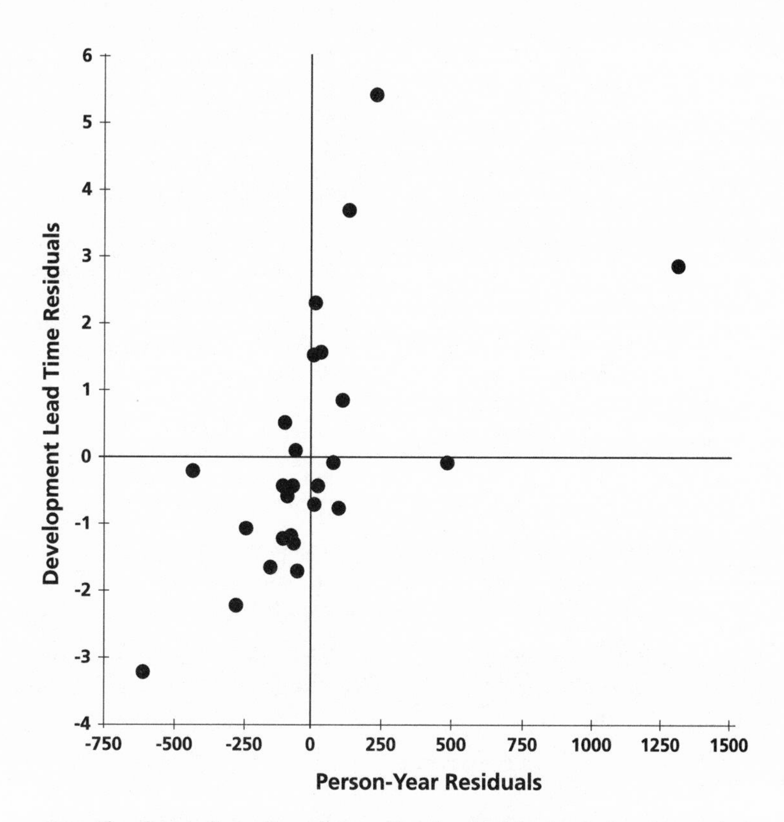

NOTE: The data indicate the relative effort involved in developing the product plotted against the relative speed, after normalizing for project content.

characteristics of the development process employed, *internal integration*, *overlapping problem solving*, and *external integration*. This discussion addresses internal integration. Internal integration estimates the extent to which the different subgroups involved in development activities are managed and coordinated to achieve a well-integrated, coherent product (Table 4-3) (see Clark and Fujimoto, 1991; and Bowen, et al., 1994). The index was constructed by adding the variables listed in the table.[7]

The index of internal integration was critical in explaining performance differentials in studies of the automobile industry (Clark and Fujimoto, 1991). Figure 4-4 shows, however, that its values do

TABLE 4-3

Indicators of Internal Integration During Development Activities

Indicator	Definition
1. Development Project Manager (PM) exists	Someone is responsible for the development activities.
2. PM responsible for wide development areas	The Project Manager has wide responsibilities, encompassing the development of the entire module.
3. PM performs product planning	The Project Manager is actively involved in product planning activities.
4. PM responsible for layout	The Project Manager is directly responsible for specifying the basic architecture of the new product.
5. PM performs concept generation	The development Project Manager is the driver in the choice of the technological concept.
6. PM influences engineering	The Project Manager exerts strong direct influence on scientists and engineers involved in the project.
7. PM has direct contact with working engineers	The Project Manager has regular direct, face-to-face contact with scientists and engineers.
8. PM has strong influence outside engineering	The Project Manager exerts strong influence over decisions made outside the engineering function, in manufacturing, for example.
9. PM has direct contact with "market"	The Project Manager has direct contact with downstream users of the product, such as mainframe system designers and engineers.
10. Development core team dedicated	A core team is dedicated to the development of the new subsystem.

The definitions are broadened here for applicability beyond the automobile industry. With internal integration, the last indicator, *development core team dedicated*, was substituted for one of the original indicators, *liaison persons have strong influence over working engineers*, since most projects covered here were too small to have formal liaisons.

SOURCE: Adapted from K. B. Clark and T. Fujimoto, *Product Development Performance: Strategy, Organization, and Management in the World Auto Industry* (Boston, MA: Harvard Business School Press, 1991).

not correlate with performance across my empirical sample. Figure 4-4 divides my projects into two groups: high-performing projects, with project resource values lower than the average, and low-performing projects with higher-than-average resources. Figure 4-4

FIGURE 4-4

Variation in Internal Integration Among High and Low Performers

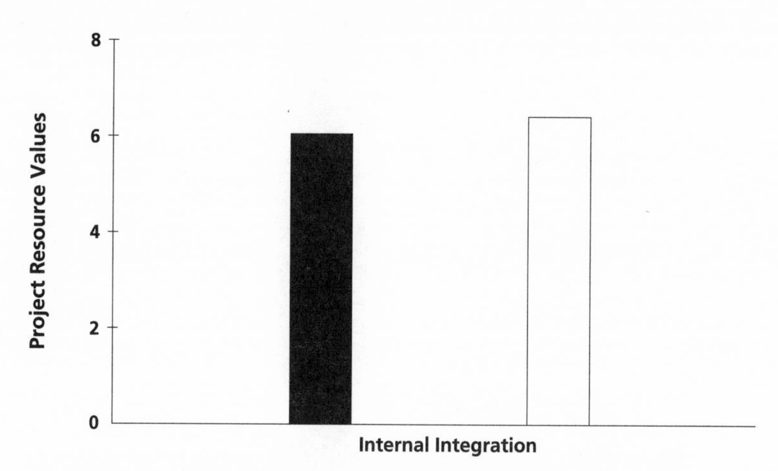

NOTE: Black bar indicates high performers; white bar indicates low performers.

reveals that the average values of internal integration are quite similar for the two groups (differences are in the right direction, but not statistically significant).

Most projects looked at, regardless of geography, had adopted several of the methodologies described in the work of Allen, Clark, and Fujimoto. They employed project-focused organizational structures for development; most of their team members were dedicated; and designers and manufacturing engineers communicated frequently. Many of the skills, practices, and methods leading to effectiveness in development had thus been employed by many competitors and no longer appeared to differentiate the best.

Table 4-4 uses regression analysis to explore this notion further. It provides more structured evidence that the indexes do not explain the vast lead time or productivity performance differences in the empirical sample. Internal integration is not significant when regressed against person-years or development lead time. The impact of other indexes on performance was also not significant (see Iansiti, 1995a-c).[8]

These results therefore indicate that the process followed in the development stage does not, by itself, explain the project performance puzzle posed above. Other traditional factors in explaining project performance differences could also be eliminated. Project performance was *not* correlated to any of the following: competitive

TABLE 4-4

Regression Results Testing the Relationship Between Internal Integration and Project Performance

Dependent Variable	Development Lead Time	Person-Years
Constant	4.0 (2.8)	443 (480)
Internal integration	0.12 (0.47)	–38.9 (81.8)
Japanese company	–1.65* (0.82)	–286* (141)
Technical content	1.81 (1.07)	673*** (186)
Adjusted R-squared	0.14	0.37
Residual degrees of freedom	23	23
F-test	2.4*	6.0****

The table displays regression coefficients unless otherwise indicated; numbers in parentheses are standard errors. A * indicates significance at the 10 percent level, ** significance at the 5 percent level, *** significance at the 1 percent level, and **** significance at the 0.1 percent level.

position in the industry (market share), resources employed, or aggregate cumulative experience (total number of product generations introduced). The number of patents obtained also did not have a significant statistical association with either development lead time or person-years.

In summary, the evidence on the impact of traditional development performance drivers is consistent with the view that most competitors had already adopted them. All competitors had high R&D investment and organizational structures and behavior that emphasized communication and integration during the development stage. These factors had become necessary to be competitive, but were not sufficient to lead. The differentiating factor among projects was instead in the upstream activities of technology integration.

Assessing Technology Integration Capability

I now turn to the technology integration process. Before its impact could be measured, however, I needed to define a methodology to assess its characteristics. During the first or pilot phase of the empirical work, the managers and technical experts interviewed identified a set of indicators as being particularly important in satisfying the three factors outlined earlier (knowledge generation, retention, and application). As with the development process variables, I defined a series of technology integration indicators (0–1 variables) which were added together to construct an overall index of technology integration capability. Table 4-5 lists the variables (see Iansiti, 1995a-c, for additional details on this methodology).

It is critical to underline the differences between the technology integration capability index and the indexes for development capability (that is, internal integration or overlapping problem solving). Not only do the indicators focus on different activities, but they refer to completely different stages of the project. The first refers to activities performed during the technology integration stage, *before* the technical concept is established. The other two refer to the development stage—by definition *after* the establishment of the technical concept. The first is therefore aimed at analyzing the earlier project stages and is (in measurement) completely independent of the process employed in later project stages—which is the focus of the other indicators.

Technology Integration and Project Performance

To investigate the association between technology integration capability and project performance, the sample was, again, divided into high- and low-performing groups. But this time the difference in the index between the two groups is quite large, going from an average of about 4.5 to about 10.5 (see Figure 4-5; the difference

TABLE 4-5A

Indicators of Technology Integration Capability (Knowledge Application)

Knowledge Application:	Mechanisms for the systemic and timely application of the knowledge generated and retained.
Integration group exists	The responsibility for integration is located within a single unit or core team. This unit is defined as the integration group.
Integration team dedicated	A core scientist/engineer team is dedicated to technology integration activities.

TABLE 4-5B

Indicators of Technology Integration Capability (Knowledge Generation)

Knowledge Generation:	Mechanisms for the generation of knowledge of the product and production system over time.
Day-to-day contact with plant	Integration group is in ongoing daily contact with manufacturing plant. Team members talk by phone or in person with plant employees on at least a weekly basis, to investigate current production issues.
Fixes production problems	Integration group is responsible for major production problems of ongoing product lines. Integration team members have been called in to fix production problems within the last six months (and thus have detailed experience with experimentation tools on production floor).
Manufacturing engineering group not on critical path	The integration group has explicit responsibility for the specification of the production process (and is, therefore, intimately familiar with it).
Relocation at source of uncertainty	Integration team moves to the major sources of technical uncertainty and performs experiments, such as pilot trials or yield improvement, at the production facility during ramp-up.
Chooses production volume equipment	Integration group has explicit responsibility for choosing production equipment (and is, therefore, intimately familiar with it).
Interacts with system group	Integration group members interact directly with mainframe system designers. There is no liaison group for transferring information between the two.
Interacts with component groups	Integration group interacts directly with component designers, such as chip developers. No liaison group acts as an information transfer between the two.

between the two groups is statistically significant). This implies a substantial difference in technology integration process between high- and low-performing projects.

A more rigorous test further supports the conjecture that technology integration is linked to performance. Table 4-6 shows several regressions. As expected, technology integration capability is associated with both development lead time and resources. The lack of significance in the Japan "dummy" variable implies that technology

TABLE 4-5C

Indicators of Technology Integration Capability (Knowledge Retention)

Knowledge Retention:	**Mechanisms for the retention of knowledge of the product and production system over time (especially through the build-up of individual project experience).**
Technical expert exists	An individual with great depth and breadth of knowledge is a member of the integration group. This is defined as having at least ten years' experience on multiple development projects aimed at high-end computer packaging.
Technical expert is project manager	The expert is the champion of the entire project, and is not relegated to an advisory role.
"T" specialization	Integration team members have deep knowledge in a specified technical area but are exposed to how this knowledge base interacts with a wide variety of activities. This is affirmative if at least 50 percent of the integration group members are focused on a given technical area (that is, cooling, lithography, and so on) but have had experience with how decisions made in that area impact all major project stages, from concept development to manufacturing.
Continuous cycle (firm)	The firm markets a continuous stream of technically related products. The current project is not the first advanced packaging module for high-end computers.
Continuous cycle (group)	The integration group works on a consistent stream of products and experiences continuity of membership. At least 30 percent of the integration group members have experience in the previous advanced packaging project.

integration capability is associated with variation in performance *within* the Japanese group of projects, as well as *within* the United States and European group. To support this point, I plotted the actual values of project performance and technology integration index for the Japanese and the U.S. and European samples of projects (see Figure 4-6). These plots are consistent with the statistical analysis, showing the general trend of high-performance projects (lower person-years and shorter development lead time) tending to be characterized by higher values of the index. I also examined the association between the three factors making up the technology

FIGURE 4-5

Variation in the Technology Integration Index Among High and Low Performers

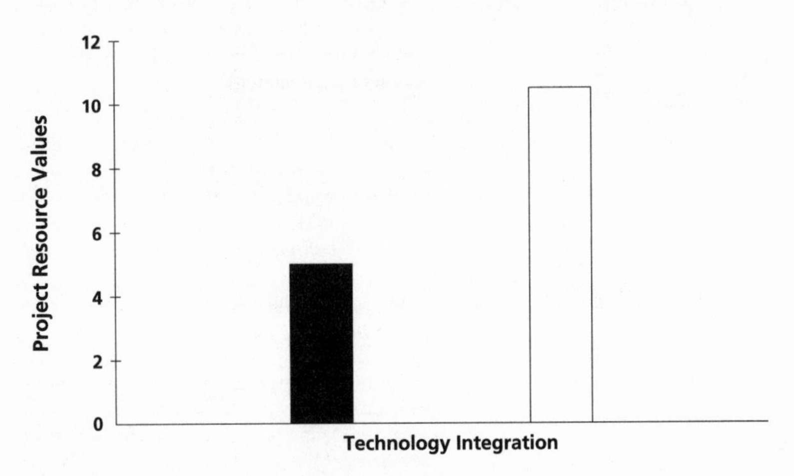

NOTE: Black bar indicates low performers; white bar indicates high performers.

TABLE 4-6

Regression Results for Performance Models Including Technology Integration

	Development Lead Time	Person-Years
Constant	6.6	542
Japanese company	–0.08 (0.79)	–73 (138)
Technology integration index	–0.28*** (0.10)	–47*** (17)
Technical content	1.57* (0.91)	605*** (160)
Adjusted R-squared	0.36	0.52
Residual degrees of freedom	23	23
F-test	6.0***	10.4***

The table displays regression coefficients unless otherwise indicated; numbers in parentheses are standard errors. A * indicates significance at the 10 percent level, ** significance at the 5 percent level, and *** significance at the 1 percent level.

integration index and performance. The results are shown in Table 4-7. The association is statistically significant for all three.

In summary, the results show that technology integration capability is associated with project performance. All three types of mechanisms identified—knowledge generation, retention, and application—were significant. Other, more traditional indicators of development performance, such as internal integration or overlapping problem solving between design and manufacturing engineering, did not appear to be as important. These results suggest that what happens before technologies are selected can have a very significant impact on the performance of the entire project, a finding consistent with the arguments made in Chapter 2.

Process and Performance in Semiconductors

Semiconductors may be *the* ideal environment for the study of technology choices. If technology integration is important to performance in the development of mainframe subsystems, in the semiconductor industry it is essential to competitive survival. In the DRAM (dynamic random-access memory) market, the investment required to build a single generation is in excess of $1 billion. Moreover, the competition is so intense that virtually all profits are made in the first year after a new generation is introduced. The scale of the manufacturing effort and the importance of timing are so extreme that organizations simply cannot afford to make mistakes in the choice of technological possibilities; a wrong choice can easily trigger a delay of many months and represent the difference between huge profits and dramatic losses. The effective integration of technology, therefore, *drives* competition.

The development of the 1 megabit DRAM has become a classic example of this phenomenon. Almost everyone in the industry decided to attack the challenging density requirements for the new generation by choosing novel technologies for the DRAM cell (either stacked or trench capacitor). Toshiba had made the earliest investigations into one of the new technologies and announced at an annual semiconductor conference that it might pursue it. Toshiba followed this announcement with several years of technology integration, however, characterized by careful experimentation to test the new cell structure in a representative manufacturing context. After additional evidence was gathered, the company decided that the new cell was too risky and opted for a simpler, more conservative technology choice. This approach turned out to be much easier

FIGURE 4-6

The Relationship Between Technology Integration and Project Performance in Different Geographical Areas

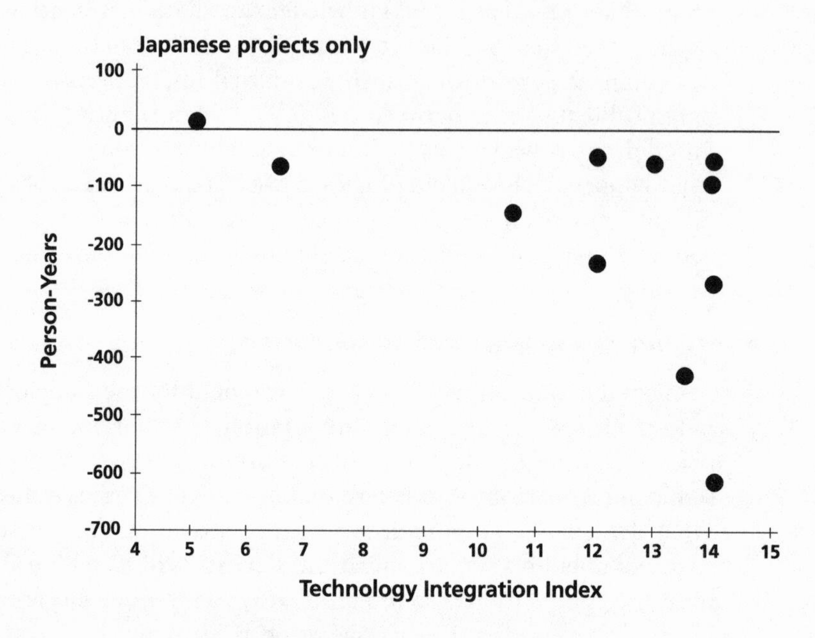

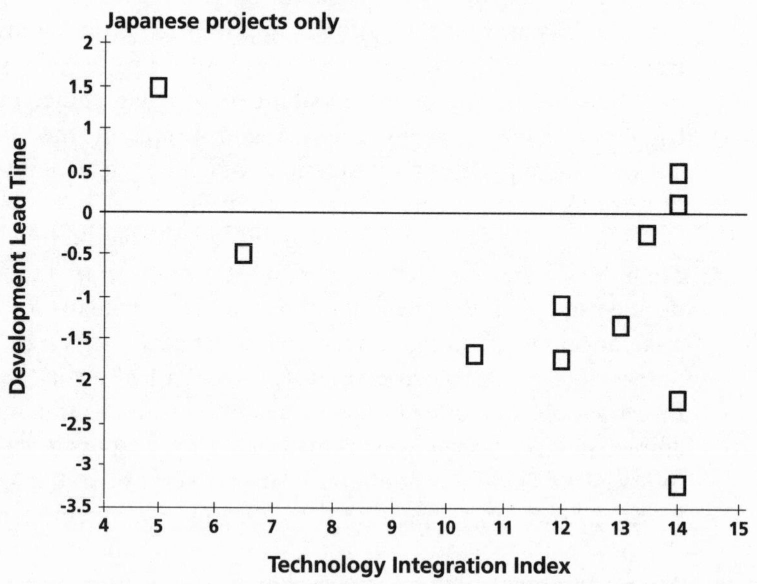

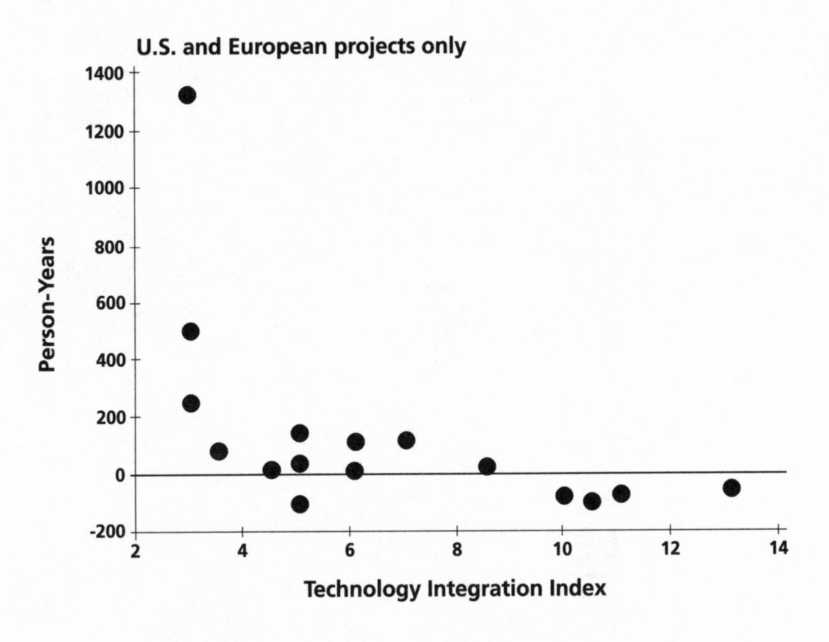
U.S. and European projects only
Person-Years
1400
1200
1000
800
600
400
200
0
-200
2
4
6
8
10
12
14
Technology Integration Index

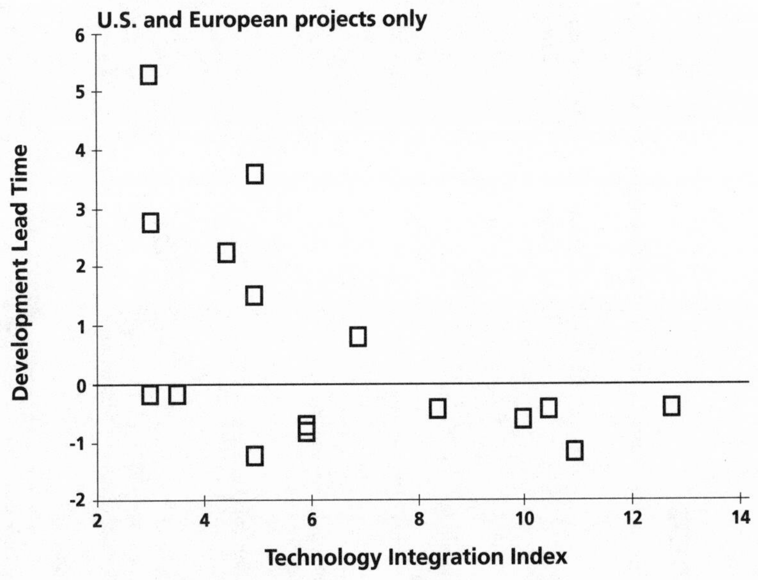
U.S. and European projects only
Development Lead Time
6
5
4
3
2
1
0
-1
-2
2
4
6
8
10
12
14
Technology Integration Index

TABLE 4-7

Regression Results for Models Breaking Down the Association of the Three Factors

	Development Lead Time	Development Lead Time	Development Lead Time	Person-Years	Person-Years	Person-Years
Constant	5.78	6.14	6.49	476	470	521
Knowledge application	–1.24** (0.45)			–278*** (72.6)		
Knowledge generation		–0.477*** (0.13)			–85.9*** (22.6)	
Knowledge retention			–0.74*** (0.24)			–129*** (42.6)
Technical content	1.67 (0.98)	1.29 (0.90)	1.92* (0.96)	614*** (160)	543*** (161)	655*** (173)
Adjusted R-squared	0.24	0.42	0.29	0.51	0.50	0.42
Residual degrees of freedom	24	24	24	24	24	24
F-test	5.25**	8.72***	6.33***	14.3***	14.1***	10.6***

The table displays regression coefficients unless otherwise indicated; numbers in parentheses are standard errors. A * indicates significance at the 10 percent level, ** significance at the 5 percent level, and *** significance at the 1 percent level.

to implement, and Toshiba's DRAMs were on the market more than a year before those of its competitors. (Most other firms had responded to Toshiba's announcement by rushing to develop new capacitor technologies, thereby shortchanging technology integration project phases.) This created fabulous profits for Toshiba and large losses for its competitors, many of whom exited the industry.

Semiconductor process development is also an ideal environment to study because many measures of performance and process are even better defined and easier to gather than they were in the mainframe study. It was therefore possible to go beyond qualitative characterizations of organizational process to measure exactly how much experimentation and experience were involved in the projects.

Empirical Environment

Before diving into the data, it is useful to describe the environment in a little more detail. My study focused on the development of major new generations of semiconductor process technology. These generations did not merely involve incremental evolution, but were characterized by rapid and discontinuous technological change. Challenged by the fundamental physical limits of the wavelength of light, projects introduced a variety of new approaches in recent years. Several authors (see, for example, Davari and Dennard, 1995) describe how technical challenges have been sharpened and broadened, as technologies move increasingly close to several fundamental limits. Deep ultraviolet lithography, for example, has replaced traditional lithography techniques. Electron beam and X-ray techniques are also in use, particularly for mask making and repair. Fundamental changes in the structure of semiconductor devices have also been implemented. Some of the more significant innovations have focused on the ability to make flat surfaces that enable three-dimensional structures (chemical-mechanical polishing, planarization, shallow trench isolation) and on capacitor design techniques for DRAMs (traditional, trench, or stacked). These technical challenges have meant that a given generational change might involve replacing 20 percent to 80 percent of production equipment and steps. Moreover, the equipment replaced is typically that most critical to product performance (for example, mask aligners and etchers).

Such changes create significant implementation challenges, since the context of application for these technologies is extremely complex. More than two-thirds of the $1 billion that a full-scale semiconductor fabrication facility ("fab") cost in 1995 was typically spent on production equipment. A semiconductor production process included hundreds of individual processing steps.

The changes are thus both discontinuous and extensive. The evolution from "i-line" to deep ultraviolet lithography (DUV), for example, which characterized several projects in our study, involves massive changes in production equipment and infrastructure. The transition to DUV entails a complete redevelopment of the lithography equipment, which is the most complex and expensive equipment in the process. The transition also completely changes the chemistry of the production process, requiring the development of new resist compounds. Finally, the transition between i-line and DUV impacts almost every step of the production process; it creates a sharp discontinuity in the relationship between fundamental technology and context of application.

At the same time, the semiconductor environment is an excellent area of study because products exhibit a well-defined set of functional features that are relatively simple to track and model. As such, it is an environment in which discontinuities in the fundamental technology are common, while product performance measures are stable enough to allow a good analytical comparison of products and projects.

The empirical focus of this study was on two specific segments of the semiconductor environment: high-volume logic (microprocessors) and memories (DRAMs). Each segment is sharply defined, with competitors facing comparable challenges. Here, only projects in each segment will be compared to projects in the same segment to make sure that a precise, appropriate, comparison is made. Memory process development projects are thus only compared to similar memory process development projects, and only logic process development projects are compared to logic process development projects. The technical data base covers all major new product introductions in the period from 1985 to 1995. The organizational database comprises a smaller sample of twenty-nine projects, including some conducted by each major firm during that same period.

Project Performance

The first step in data interpretation is to look at performance, which in semiconductors is characterized by different priorities than those prevailing in mainframe subsystems. The most important success factor for semiconductors is being on the market first with a new process generation. The new generation must also have a small "die" (silicon chip) size, packing transistors together as densely as possible, since density is the major driver of both cost and product performance. The central measure of project performance is represented by the transistor (or gate) density. Since average density changes rapidly with time, as described by Moore's law, the measure is "detrended" for this study, using the residuals of a regression of the logarithm of density against time. This measures how far ahead (or behind) the industry trend a given project is.

Table 4-8 provides the foundation for this calculation of project performance, showing regressions between the logarithm of gate density and time for all of the products included in the sample. The analysis is performed separately for memory and logic products. The table shows that the trend is for gate density to increase very rapidly with time (more than ten times in ten years for DRAMs and slightly less for logic). The residuals from these regressions form the basis for my measure of project performance.

Tables 4-9 and 4-10 analyze the differences between the major competitors in DRAMs and microprocessors. Each includes three columns. The first displays the detrended gate density, which is the average residual for each company in the sample. A negative value implies that the company is, on average, behind the trend line; a positive value implies that it is ahead. The next column, labeled time vs. trend line, translates these measures in a year value, illustrating how many years a given company is in front of or behind the trend. Finally, the third column shows the actions the firm has recently taken to respond to its competitive situation.

TABLE 4-8

The Evolution of Transistor Density for Memory and Logic Circuits

Dependent Variable	Log of Transistor Density (Memory)	Log of Transistor Density (Logic)
Time	.141***	.082***
	(.005)	(.007)
R-squared	.93	.59
F-test	845.9***	135.2***
N	65	94

The table shows regressions of gate density against time. A * indicates a 10 percent level of significance, ** indicates a 5 percent level, and *** indicates a 1 percent level.

TABLE 4-9

Differences in Performance Averages Among Major DRAM Firms

DRAM Firm	Detrended Log (Transistor Density)	Time vs. Trend Line (Years)	Actions
INMOS	–0.30	–2.10	exit
Motorola	–0.28	–1.96	exit
AMD	–0.27	–1.93	exit
Mostek	–0.23	–1.62	exit
National	–0.19	–1.33	exit
Intel	–0.15	–1.09	exit
Micron	–0.11	–0.75	
Mitsubishi	–0.10	–0.68	
Fujitsu	0.03	0.18	
AT&T	0.04	0.28	exit
TI	0.06	0.41	
IBM	0.07	0.47	
Toshiba	0.07	0.51	
NEC	0.14	0.99	
Hitachi	0.23	1.61	
Samsung	0.26	1.83	

The table displays the differences in detrended transistor density and then translates these into year equivalents. A time versus trend line value of *1*, for example, implies that the organization is, on average, one year ahead of the transistor density trend in the industry. The action column lists any major actions taken by the organization in the last ten years.

The tables highlight very substantial differences among competitors, which in several instances build up to many years. Such time lags are essential to competitive survival in this environment. The DRAM table shows that companies that were behind exited the industry. In the microprocessor environment, process development is not the only means of differentiation, since the products are not virtually identical commodities. As a result, rather than exiting

TABLE 4-10

Differences in Performance Averages Among Major Microprocessor Firms

Microprocessor Firm	Detrended Log (Gate Density)	Time vs. Trend Line (Years)	Actions
HP	–0.37	–4.55	Access Intel process
Sun (accessing Fujitsu)	–0.25	–3.11	Access TI process
Cyrix	–0.19	–2.35	Access IBM process
Motorola	–0.08	–0.94	Access IBM process
AMD	–0.04	–0.46	Merge w/NexGen
MIPS (accessing NEC, Toshiba)	0.05	0.64	
DEC	0.06	0.69	
Intel	0.17	2.10	
NexGen (accessing IBM)	0.19	2.32	
IBM	0.20	2.39	

The table displays differences in detrended transistor density and then translates these into year equivalents. A time versus trend line value of *1*, for example, implies that the organization is, on average, one year ahead of the transistor density trend in the industry. The action column lists any major actions taken by the organization in the last ten years.

entirely, firms sought alliances with competitors so as to focus internally on design while accessing external process development capabilities. These observations confirm that gate density is indeed critical to competitiveness.

Assessing the Process

The study's first phase was an eight-month period of exploratory analysis, which involved discussions with technical and industry experts and focused on extensive informal investigation of one major U.S. organization. These investigations were used to refine the basic conceptual and empirical frameworks for the study and to develop both a structure for future interviews and a detailed questionnaire.

Observations on major recent projects performed by most major organizations in the semiconductor industry (their combined market share was over 80 percent) were then collected. The database covers the introduction of twenty-nine major new process technology generations.[9] Table 4-11 includes a list of some of the more basic variables measured.

As with the mainframe study, technology integration variables were divided into three groups, each associated with one of the three types of mechanisms (knowledge generation, retention, and application). Table 4-12a shows the variables associated with the knowledge application mechanism, which replicate the definitions for the mainframe study. Tables 4-12b and 4-12c show the variables used to examine the impact of the other mechanisms. As noted, quantitative indicators were more central than they had been in the mainframe study. The variables are generally not *0-1* indices but estimates of actual physical quantities, such as the capacity of the experimentation facility used. This allowed a deeper understanding of the technology integration process.[10]

Technology Integration and Project Performance

The analysis once again started with the examination of the extent to which more traditional explanations of R&D performance could

TABLE 4-11

Basic Project-Level Variables

Variable	Definition
Sales	Semiconductor sales, used as a proxy for organization size.
Capital investment ($MM)	Overall investment in semiconductor capital equipment.
R&D spending ($MM)	Overall investment in the development of semiconductor process generations.*
R&D human resources (PY)	Number of person-years of effort used by the project. This includes all functions, internal and external to the corporation, involved in the introduction of a new process technology generation.

*These numbers were fairly representative, since the projects covered in the study were *the* major process development activitiy during the time of the project. Information on investment at the project level was also gathered, but the data were much less reliable, largely because of differences in accounting systems.

be used to interpret results. As with the mainframe study, I found that that differences in R&D performance could not be explained by broad, firm-level variables or by traditional performance drivers (see Iansiti, 1997; and West, 1996). The explanation must therefore reside in other areas, such as technology integration.

Table 4-13 displays a regression model investigating the relationship between the performance of a technological generation and the process followed for technology integration. It shows that mechanisms for experimentation and experience-building employed during technology integration are associated with performance. Specifically, project experience, research experience, experimentation capacity, and experimental iteration time are all correlated with performance in the expected way.[11]

Separating the impact of the three types of mechanisms identified above benefits the discussion of results.

TABLE 4-12A

Knowledge Application Variables

Variable	Definition
Integration group exists	The responsibility for integration is located within a single unit or core team. This unit is defined as the integration group.
Integration team dedicated	A core scientist/engineer team is dedicated to technology integration activities.

TABLE 4-12B

Knowledge Generation Variables

Variable	Definition
Experimental capacity (wafers/week)	Capacity of experimental fabrication facility used in the technology integration phase of the development project.
Experimental iteration time (weeks)	Minimum time taken to fabricate a test sample at the experimental fabrication facility.
Experimental representativeness	Fraction of experimental fab equipment (by cost) that is the same as the equipment used in the final production facility.

TABLE 4-12C

Knowledge Retention Variables

Variable	Definition
No previous project experience	Fraction of project members participating in technology integration decisions that had not participated in a complete previous process technology introduction project in the same environment.
1 full generation experience	Fraction of project members participating in technology integration decisions that had participated in only one complete previous process technology introduction project in the same environment.
2 full generations experience	Fraction of project members participating in technology integration decisions that had participated in only two complete previous process technology introduction projects in the same environment.
>2 full generations experience	Fraction of project members participating in technology integration decisions that had participated in more than two complete previous process technology introduction projects in the same environment.
Research experience (internal)	This variable is a one if at least one of the project members involved in the technology integration decisions had spent more than three years working on a relevant technology in the research organization.
Research experience (university)	Fraction of project members having a Ph.D. and research experience in a related technology at a university.

Knowledge Application

The knowledge application factor was not found to be useful in distinguishing between the best and worst performers. All projects had satisfied the basic criteria for both index variables. With no exceptions, all had dedicated resources to technology integration, attempting to make critical choices in a holistic fashion. At the time of this study, in the semiconductor environment, an organizational focus on technology integration had become second nature—a necessary but not sufficient requirement to be competitive. One could argue, in fact, that almost all activities performed by the development group in the semiconductor industry are aimed at technology integration—many activities defined as traditional development have been subcontracted to suppliers or handed off to plant personnel. Among these, the most critical is probably equipment design

TABLE 4-13

Determinants of Integrated Circuit Performance (Transistor Density)

Dependent Variable	Logarithm of Transistor Density
Intercept	–273*** (19)
Time	0.138*** (0.010)
DRAM	0.775*** (0.083)
R&D resources (PY)	–0.0005* (0.00026)
Experimental capacity (wafers/week)	0.00019*** (0.000059)
Experimental iteration time (weeks)	–0.032** (0.014)
No previous project experience	–0.372* (0.203)
Research experience	0.115* (0.067)
F test	37.469***
R-squared	0.926
N	29

The dependent variable is the logarithm of the transistor density. Time is a continuous variable corresponding to the time of process introduction. The DRAM dummy is *1* if the project was aimed at the DRAM segment, and *0* otherwise. A * means significance at 10 percent level, ** significance at 5 percent level, *** significance at 1 percent level.

and development, which has now been almost entirely taken over by an active supplier base.

Technology integration is thus the central activity in process introduction projects in semiconductors. For members of the development team, most daily tasks are aimed at testing the capabilities of new equipment, with the goal of integrating each of the modules into a coherent production process capable of producing high-density circuits with high yield. As a senior manager at Intel said, "The integration of technologies is our most central capability, and it has always been."[12]

While no gross distinctions among projects were found, however, there may be indications of more subtle factors. The consistently negative association between R&D resources and performance, for example, may be caused by more that just coordination problems. It may suggest that smaller groups find it more difficult to integrate the knowledge generated in projects and apply it in a holistic fashion. This implies that smaller, more tightly integrated groups of experts should be more effective at integrating novel technologies.

Knowledge Generation

Although all competitors found experimentation to be important in making good technology choices, I found a large variation in the actual capability of competitors to experiment. The most striking differences were in experimentation capacity, that is, the number of experiments an organization could perform in a given period. The observed variation among projects was more than a factor of ten, indicating fundamental differences in approach. The amount could be truly enormous: Some projects would run *millions* of trials during a project, investigating a broad set of technical alternatives. Also significant was the speed of experimental iterations, given by the shortest time in which a test sample could be run.

I found that the nature of experimentation capabilities was closely associated with process performance. Competitors that could run many accurate trials could make superior technology choices and bring their new processes to market faster. The results, therefore, indicate that experimentation capability can be critical to the rapid discovery of which technological choices work best in a complex application environment.

Knowledge Retention

I also found a substantial variation in research and project experience among projects. Some organizations relied almost entirely on a small group of highly experienced individuals to make technical choices. These engineers had generally introduced a large number of project generations before and were intimately familiar with the details of their production setups. Other organizations appeared to rely more on recent recruits, either from leading Ph.D. programs or from the research organization.

Overall, both research and project experience positively contributed to performance. Some level of project turnover was found to be useful, however. By and large, projects staffed mainly by members with more than two full generations of experience did not perform as well as those with some lesser amount of experience. Several project members elaborated on this point, stressing the importance of bringing in employees with new experience and perspectives. One explained it this way:

> We try to bring in about 10 percent new project members every year. This achieves two things. First, we get enough new people trained so that there is a solid base for future projects. Second, it brings in new ideas and makes sure we don't get too inbred.[13]

New project members were frequently chosen to bring critical research experience from universities, said another study member:

> There are several university departments that we monitor very closely. We know what students are working on, and when they are graduating. We will choose a few of these students that have worked on technologies that are important to us and target them for hire. Hiring these hand-picked graduates is a critical source of new knowledge for us.[14]

This point will be expanded upon in Chapter 7, when different technology integration strategies are distinguished.

In summary, the study results show significant differences among projects in the ability to introduce new process technology generations in a timely and effective fashion. Technology integration is critical in differentiating between the best and worst projects. Variables characterizing differences in experimentation capability and in research and project experience were significantly associated with differences in project performance. As with the mainframe data, differentials in performance are not associated with the development of given technologies, but with the processes aimed at choosing what gets developed in the first place.

Process and Performance in Workstations and Servers

In semiconductors, technology integration efforts are challenged by the application of uncertain hardware technologies (primarily new materials and equipment) to enormously complex and expensive manufacturing environments. The study of workstations and servers is useful in characterizing the linkage between technology integration and performance in a very different setting. Here, the complexity is not in the manufacturing environment, but in the architecture and use environment of the product. Additionally, the performance drivers in workstations are also different from those in semiconductors. As with semiconductors, speed is essential—the products must reach market quickly to maximize the performance advantage of new component technologies. Flexibility is also needed, however—projects must be sensitive to rapidly evolving market requirements, changing project specifications multiple times to make sure that the final product really meets the current needs of its fickle customer base. This is necessary because, unlike microprocessors and DRAMs, market requirements for workstations will evolve multiple times during the course of a single project. (This phenomenon is explored in Chapter 9.)

As with the mainframe and semiconductor studies, the study of technology integration in workstations and servers involved visits to most of the competition in the industry, including Sun, Silicon Graphics, IBM, and Digital Equipment Corporation. (For details of the study, as well as for more quantitative performance and process measures, see Iansiti, 1995d; and Iansiti and MacCormack 1997). Once again, my research detected very large differences in the performance of different projects along all critical dimensions of performance—total lead time, development lead time, and person-years (see Table 4-14).

The differences are exemplified by Figures 4-7a and 4-7b, which display the time lines of two projects. The first was the development of Silicon Graphics' Challenge, one of the most effective projects in our sample; the second, which will remain anonymous, was completed by one of SGI's competitors. Both projects were aimed at developing a very similar system; architecture, performance, and overall system complexity were very close. The performance of each project was dramatically different, however: The Challenge was more than one year faster—an eternity in this business. The difference was due to an enormous variation in the development lead time (defined as the time between concept freeze and launch). The development lead time was only five months for the Challenge, compared to a year and a half for the other project. The person-year resources were comparable for the two projects, as was the structure of the project team. Both projects were staffed by strong and influential project leaders, and both had a similar organization. The Challenge, however, approached technical decision making in a very different way. First, it employed a much deeper experience base in the core decision-making team. Second, it ran a more extensive experimentation effort, characterized by higher

TABLE 4-14

Variation in Workstation Project Performance

	Maximum	Minimum
Total lead time (months)	42	27
Development lead time (months)	20	5
Person-years of activity	600	70

The table is drawn from data on nine comparable server projects.

simulation capacity, a higher diversity of simulation tools, and more extensive use of partial prototypes and mock-ups. By the time the concept was frozen, SGI engineers had a much better view of how the whole system fit together and functioned. As with mainframes and semiconductors, the difference was in technology integration.

The evidence indicates that effectiveness in workstations and servers also hinges on the capability for technology integration. While the nature of the technologies and of the context are different from those found in semiconductor process development, the nature of the problem is the same: choosing novel technologies in a complex environment. What made workstation manufacturers successful were the same fundamental factors that explained the effectiveness of certain projects in the semiconductor and mainframe studies: knowledge generation, retention, and application.

Knowledge Application

In the most effective workstation and server projects, critical decisions were made rapidly and jointly by a dedicated core business team, usually a group of professionals who met daily. The core of the business team comprised a group of architects, who had the final say in integration decisions linking technological possibilities and system architecture. In the Challenge project at Silicon Graphics, this group included both newcomers and SGI veterans. In all cases, architects had both extensive experience with multiple aspects of workstation and server design and the capability to make decisions, with a good understanding of their impact on system-level performance.

Differences in the coherence of the technology choice process were larger than those found in the semiconductor industry. This was caused by frequent division of responsibilities between software and hardware architects. In some firms, technical decisions

FIGURE 4-7

a. SGI's Challenge Project

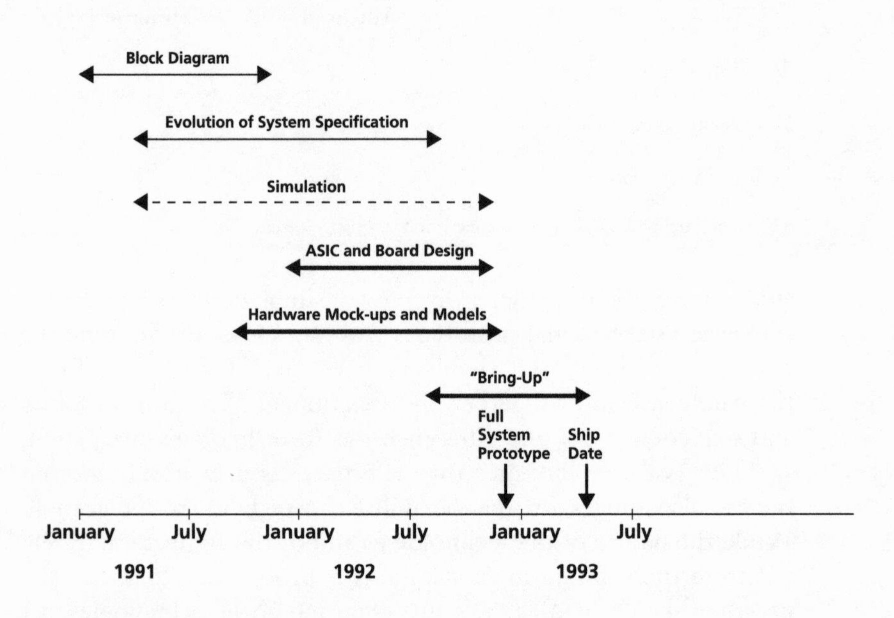

b. A Competitive Project

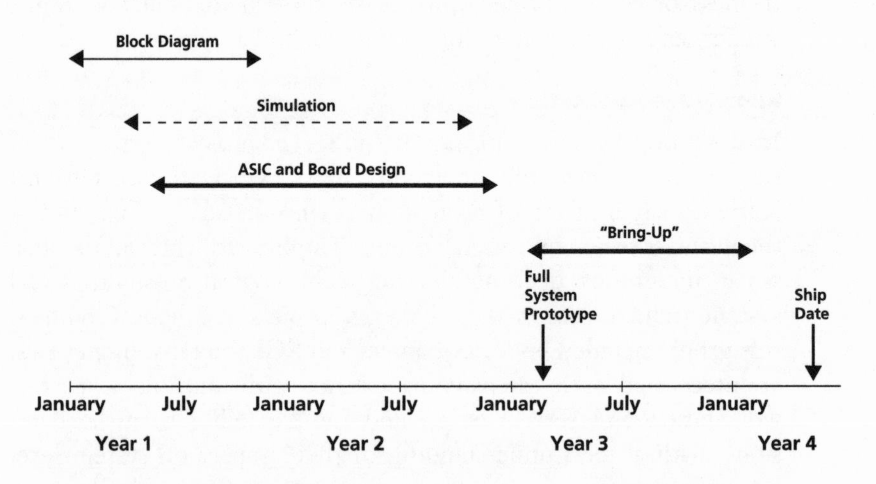

NOTE: Conceptualization activities are indicated by the thin arrows, implementation activities by the thick arrows. Simulation activities are indicated by broken arrows to show that they were used both for concept evaluation and for testing detailed designs.

SOURCE: Adapted from Ellen Stein and Marco Iansiti, "Silicon Graphics, Inc.," Case 9-695-061 (Boston: Harvard Business School, 1995).

regarding software and hardware were handled by separate organizations, which did not always communicate effectively. In the most effective projects, like SGI's Challenge, the two organizations were instead managed as a coherent group, and architects and engineers worked together daily.

These organizational challenges were aided by information technology. In most projects, as the effort progressed, the specification changed from a paper document to the actual software code representing the product design for computer simulations. This source code performed the function of a living specification; it was a complete representation of the electronics in the product, which all members of the project team shared. The source code enabled the system-level impact of individual technical choices to be rapidly verified. It also facilitated communication among team members, and it integrated individual efforts. All team members could work on the same software model of the product's design. At Silicon Graphics, this code was kept in a master library, and when a significant change took place, each key member was automatically notified via electronic mail. As such, everyone was apprised of the latest version of the workstation system and the direction of its evolution. Still, information technology did not appear to replace the need for constant human interaction.

Knowledge Generation

Rather than the massive development fabs of semiconductor process development, knowledge generation in workstations is driven largely by computer simulation. Typically, a project team can already run system performance simulations after only a few months of project work. Simulations will encompass the entire system electronics, from the ASIC logic at the chip level to the design of the circuit board. The programs can be linked so that the entire computer system can be simulated as a single unit. In some cases in the study sample, engineers could even test parts of the actual operating system to detect problems with logic timing and performance.

While accurate, the simulation programs are relatively slow. Hardware designers and testing engineers thus also build numerous physical models of the system. These prototypes do not typically represent the entire product, but include only some critical parts of the computer system so that specific problem areas can be tested. Combined with the simulations, such partial physical prototypes help uncover problems before committing to a very expensive complete and representative prototype. These experimentation methodologies allow for very rapid iteration, linking individual design choices to the performance of the entire system and facilitating their joint optimization before the full system prototype is built. In the Challenge project, the first full system prototype was built at the end of 1992, only about three and a half months before the project's end.

Workstation and server projects also rely on several external sources to test the products and search for relevant new information during the course of the effort. First among these are its lead customers. At several companies, for example, team members repeatedly invited key customers, under nondisclosure agreements, to evaluate their progress, discuss the product in development, and try their latest software on it.

Knowledge Retention

Individual experience is, once again, essential to project performance. Many sources of knowledge need to be integrated by the team members, who make use of their experience to interpret the impact of potential changes. Project architects have great depth and breadth of experience and are particularly crucial. In the SGI Challenge project, each architect had worked on a number of supercomputer or server development projects before (at least three or four product generations). The group comprised primarily design engineers, although it also included individuals with extensive testing and manufacturing engineering experience. While their broad experience was reflected throughout the project, the earliest and latest stages were possibly the most critical.

One area where experience appeared to have the greatest impact was the early partitioning of project tasks. When a project starts, the architects first focus on a comprehensive block diagram of the system, which serves as a project blueprint. The block diagram is not used as a rigid specification, and almost all of the detailed product objectives change repeatedly. Its main role is to identify the basic project modules and to highlight the most critical interactions among them. This is essential, since the block diagram identifies what is likely to change and what is likely to remain the same. It is therefore used as an early road map, identifying the most appropriate ways to partition project tasks to minimize the complexity of implementing future changes.

This need for breadth of experience in the development process is reflected in career paths. The Challenge project leader described his approach:

I think it's best to start off inexperienced hardware engineers on simulation tasks, since these give you the best view of the entire system. As they build experience in that, we will shift them over to the design of components, such as the ASICs. Balancing the experience gathered by our engineers during their careers is probably our most crucial task.[15]

SGI's approach to hiring and career development reflected the need to build on individual experience. The firm continually attempted to hire the best available talent in the industry; it also aimed to retain critical individuals. Individual turnover was among the lowest in Silicon Valley. These practices helped to build a deep base of system knowledge dedicated to SGI's products, testing, production, and user environment, thus providing a constant, solid foundation for its flexible, turbulent development process.

Moreover, many organizations rely on ties with research institutions and universities for bringing in outside research experience. SGI's relationship with the Electrical Engineering Department at Stanford is particularly close: Ph.D. students with relevant theses are frequently hired into the company. One such group of Ph.D.s recently had an enormous impact on the design of Lego, the first distributed shared memory (DSM) server to make it to high-volume production (the project is described in Chapter 6). Colleagues in other Silicon Valley firms are another critical source of external experience; as one project member explained, "There is basically a group of twenty to thirty designers around the Valley that really understands this stuff; we all know each other well and constantly rely on each other for the latest news."[16]

Technology Integration and Project Performance

Results from three empirical environments each characterized by the combined challenges of technological novelty and environment complexity show that the process for technology integration is significantly associated with performance, often eclipsing the impact of more traditional factors such as project management or cross-functional integration. Despite this evidence, however, two basic questions remain: How could performance differences among projects be so very large? Why do they appear to be even larger than those in other, more traditional environments?

The key to this puzzle lies in the combination of uncertainty and complexity itself. If the empirical environment is mature, such differences in project performance as resources needed or lead time will depend only on how the project is executed, since choosing the technical approach does not constitute a major challenge. The potential for mistakes and inefficiencies is relatively small, since all one can do is mismanage the implementation phase, running into coordination and integration problems. If the environment is characterized by high technological uncertainty, however, making technology choices becomes a major challenge. In this kind of environment, technical choice can drive vast differences in project performance. If the technology choice is well matched to the application context, the project will be fast and efficient. If it is not, the project will be slow and inefficient, no matter what happens in the development phase. If the choice of technology is extremely bad, the project might even be impossible to complete.

Proving this point involves examining the nature of the technical choices made by organizations. If the explanation is correct, the technical choices made by high and low performers should be systematically different. This is the topic of the next chapter, which focuses on technology choice and product outcome.

Notes

1. Interview conducted by author during field research from 1990 through 1995.
2. Differences between country groups in development lead time and person-years are significant, at the 5 percent level in a *t* test.
3. The results of the two regressions were as follows: *Person-Years = 112 + 598*Technical Content (199)*, and *Development Lead Time = 4.14 + 1.59*Technical Content (1.11)*, where numbers in parentheses are the standard errors.
4. A simple regression between the two residual variables is significant at the 1 percent level (*development lead time residual = .00318 * (person-year lead-time residual)*, with a standard error of *.00092. F = 12; p = 0.0019*).
5. In other words, the projects can be thought of as being on different time-cost frontiers.
6. This involved discussions with a wide variety of project participants, from scientists to development project leaders, from design engineers to tool designers, from general managers to manufacturing engineers. Although some of these interviews were structured, following a fixed questionnaire, others were free-form. Direct observations were complemented by a survey and by extensive analysis of internal documents and public sources. The characteristics of the three major project building blocks were investigated: research, technology integration, and development. Data compiled covered a variety of factors, including the skills of critical decision makers, the organizational structure, the tools and experimentation infrastructure available, and the influence of the project leader.
7. In investigating the association between technology integration capability and total and concept lead time, no significant association was found for concept lead time. This suggests that other factors may be related to the total length of

the exploration stage, such as research strategy or contacts with universities. This lack of significant association is not the result of a negative correlation between concept lead time and development lead time. The correlation between the two is weak and positive ($r = 0.195$). Total lead time is weakly associated with technology integration capability, perhaps the result of the latter's significant association with development lead time (*total lead time = concept lead time + development lead time*).

8. The only exception was overlapping problem solving, which was significant, but only in models for development lead time.
9. Both structured and unstructured interviews were conducted with scientists, engineers, and managers at different hierarchical levels in the organization who were involved with the most critical aspects of each project. The interviews comprised three phases, each conducted on separate visits. During the first phase, informal meetings with project members allowed discussion of the objectives of the study and the compilation of broad information on the nature of the company's R&D process. Before the second visit, the participants received a copy of a detailed questionnaire, which included questions on the nature of the strategy and organizational processes followed in the project, on experimentation and prototyping capabilities (such as the characteristics of the development fab used), and on the experience base of the individuals making technology integration decisions. It also captured the history of each development effort, tracking the completion of major steps as well as the resources used. Finally, it included questions about the basic characteristics of the organizations, the processes employed, and the behavior patterns of the managers and engineers. The second visit to the company focused on discussing the questionnaire, resolving any questions that managers and engineers might have, and on cross-checking the validity of the more critical information by interviewing separately different cross-sections of project participants. The third visit began with a presentation of preliminary results, followed by an extensive discussion of findings. The emphasis was on validating the accuracy of the company-specific information gathered. When possible, the interviews were recorded.
10. Several of the experimentation and experience variables are correlated. Experimental capacity is correlated with experimental representativeness (coefficient equal to *0.53*). This is not surprising, since production facilities are larger than experimental facilities. The larger the experimental facility, the more likely it is that it will employ exactly the same equipment as the final high-volume manufacturing facility. Additionally, it is also not surprising that several of the experience variables are correlated. To avoid problems caused by these correlations, I use only four critical process variables in the regression analysis presented: experimentation capacity, experimental iteration time, no project experience, and research experience (internal).
11. The regression also includes variables on R&D human resources, to normalize for the different levels observed in each project. The negative result is consistent with the observations reported about mainframes. Not all variables describing experience or experimental capability are included in this regression, to avoid correlations (see note 9). This regression includes the time and DRAM variables explicitly. Performing the regressions with the residuals from Table 4-8 as the dependent variable does not appreciably change the overall results.
12. Interview by author conducted during field research from 1990 through 1995.
13. Ibid.
14. Ibid.
15. Ibid.
16. Ibid.

The Drivers of

chapter 5

Product Performance

THE CAPABILITY FOR technology integration should be reflected in integrated products, characterized by a good match between technology and context. The investigation of this claim is the subject of this chapter. The work fits within a more general debate, however. All of us have speculated that organizations reflect themselves in the products they design. At the academic level, several authors (for example, Clark, 1985; Henderson and Clark, 1990; Tushman and Rosenkopf, 1992; and Christensen and Rosenbloom, 1995) have argued that the structure of organizations and of the products they develop converge over time. In assembled products, for example, Clark argued that functional departments come to reflect product modules, such as optics and mechanical systems.

This convergence of product and organization is a central conjecture in the field of technology management. But is it actually true at a detailed, microscopic level? Do products *really* reflect the nature of the organization that created them? This would have significant implications. It would mean, for example, that we could learn about the way Silicon Graphics manages product development by analyzing the products it creates. While arguments abound, measurements of the relationship between product and organization are rare. The reasons are fundamental to the problem. Measuring whether an organization is reflected in a product requires the simultaneous measurement of two very different types of systems: organizational and physical. On one side, we must measure the organization's characteristics, using methodologies based in the social sciences; on the other side, we must assess the product's characteristics, which requires methods based in the physical sciences. Approaching this problem therefore requires the integration of traditionally separate academic disciplines.

An empirical test of this problem is basic to this research. Developing a product demands matching fundamental knowledge (that pertaining to a particular knowledge domain) with knowledge of the specific context of application. Moreover, as argued, an effective technology integration process should lead to a better match between the two. To put these arguments on solid ground, this hypothesis should be tested empirically. Doing so requires measuring and testing the association between two separate types of factors: the characteristics of the organizational process of technology integration, and the quality of the match between technology and context in the product.

The chapter begins with a detailed discussion of two heuristic examples, followed by the development of a simple methodology that divides system-level product performance into two quantities: technological potential and technological yield. The first estimates the maximum product performance given the technologies chosen and is based on actual physical models of the product. The second estimates how much of this potential is realized in actual performance. Technological potential should therefore be linked to an understanding of fundamental knowledge domains. Technological yield, for its part, measures the quality of the match between fundamental domains and application context. Potential and yield should thus be driven by very different types of organizational processes. The methodology will therefore allow me to test the association between organizational process and characteristics of the resulting product.

Product Performance, Technology, and Integration

What determines the performance of a product? Its technological foundations certainly play a major role. In a rapidly changing technological environment, the organization's ability to include the latest technologies in its products will drive their performance. Technology is not everything, however, as Table 5-1 indicates. It compares several leading workstations on the market at the same time in late 1992. The microprocessor is the engine of a workstation, the central driver of performance at the component level. The table shows that Digital Equipment Corporation's Alpha processor is the fastest engine, with an impressive clock rate of 200 MHz, the fastest in the industry.

The data in the table challenges the expectation that the fastest components lead to the fastest product. Performance in workstations is benchmarked using a measure called *SPECmark*, which measures how fast the system executes a set of standard programs, representative of typical applications. If microprocessor clock rate is analogous to engine horsepower, the SPECmark measure is more like the performance of the automobile on a test drive. The SPECmark figures show that the highest-performance workstation included in the table is SGI's IRIS 4D/480, which also commands the highest price. Figure 5-1 shows the ratio of SPECmark performance

TABLE 5-1

Comparing Workstation Performance at the System and Component Level

Company	SGI	Sun	IBM	DEC	HP
Product	IRIS 4D/480	690MP	Pserver 560	5820	9000-750
Microprocessor chip	R3000	SPARC	POWER	Alpha	PA-RISC
Internal microprocessor clock rate (MHz)	40	64	50	200	100
Workstation performance (SPECmark)	166	91	89.3	39	77
Price	$164,900	$92,000	$74,100	$145,000	$71,000

The table shows a comparison of several competing high-performance workstations on the market in 1992. All systems were based on RISC (reduced instruction set computer) architectures and included the same number of microprocessors. The SPECmark performance measure is based on the speed at which the computers can run several important application programs. The higher the measure, the better the performance.

SOURCE: Ellen Stein and Marco Iansiti, "Silicon Graphics, Inc.," Case 9-695-061 (Boston: Harvard Business School, 1995).

FIGURE 5-1

Ratios of Workstation System-to-Component Performance

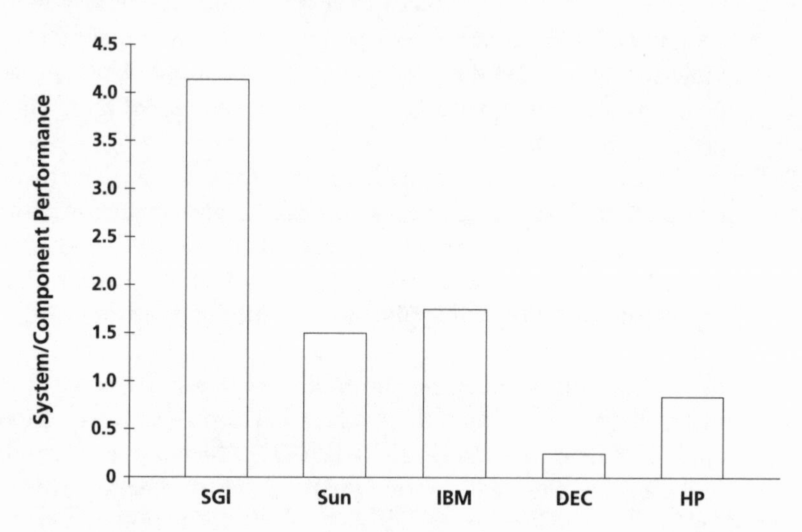

SOURCE: Ellen Stein and Marco Iansiti, "Silicon Graphics, Inc.," Case 9-695-061 (Boston: Harvard Business School, 1995).

to microprocessor clock rate, showing SGI's IRIS leveraging its component technologies better than any of its competitors.

How can the IRIS system dominate performance in its industry segment with the slowest microprocessor? Because there are multiple paths linking technology to performance in a system as complex as a workstation or an automobile. One path follows the heuristic that the best technology leads to the best product. Other paths, however, emphasize that technology needs to be matched to a complex context to work well. In a workstation, for example, microprocessors need to have an appropriately sized "bus," which feeds them with information. They must have fast access to nearby memory so that work in process can be stored and fetched efficiently. Their cycles should be matched with the input/output system, making sure that they are not always waiting for data. IRIS dominated its segment because the organization that developed it really leveraged the limited power of the R3000 by matching it to the system around it.

A complex product is not simply the sum of its components. Its design can benefit from careful optimization, working on the subtle interactions among its many subsystems. Part of system performance will scale with fundamental technology, just as a more powerful engine, everything else being constant, will lead to a faster car. But part of system performance will also hinge on how the technologies are integrated with each other. The performance of an automobile on a test drive will be a function of the interaction of its engine, brake systems, suspensions, weight, and a million other design details.

The above argument is critical to examining the relationship between organizational process and product characteristics, because the creation of fundamental technology and its implementation in a coherent product system should have very different organizational roots. While the first ought to be driven by organizational processes for deep, specialized knowledge-building, the second ought to be associated with the organizational capability for integration. This would mean that the IRIS is faster than the Alpha system because its design process was characterized by a higher level of integration.

Two Contrasting Product Systems

Table 5-2 carries the argument to a deeper level. It compares the characteristics of two mainframe processor modules encountered in the mainframe R&D study: the Z1000 and the SS180.[1] The two mainframes have very similar architectures and run compatible software, so the performance comparisons are accurate and representative.

The table displays an exhaustive set of indicators for the most important technologies integrated into the product. It also indicates existing rules of thumb for evaluating the technologies on their own merits. Finally, it displays the critical system-level measure of performance—the proportion of the cycle time that is made up of delays caused by the module. The smaller the cycle time, the faster the computer. The table is striking because the Z1000 is better than the SS180 on every single technological dimension. At the system level, however, its performance is worse. This demonstrates that a product is not just the sum of its technologies. The integration of its technical elements into a system that functions effectively in the real world is a major challenge.

The question of what makes the SS180 function faster than the Z1000 is not easily answered. It is the aggregate effect of a multitude of decisions, tricks, trials, and refinements made to optimize the system-level impact of fundamental technologies. Many of these relate to the interaction between module development and integrated

TABLE 5-2

Technology Choice and Performance in Two Mainframe Processors

Technology Variables	Z1000	SS180	Comments
Dielectric constant of ceramic	5.2	5.9	Smaller is better
Resistivity of metal on ceramic	3.5	11	Smaller is better
Dielectric constant of polymer film	3.5	3.5	Smaller is better
Resistivity of metal on polymer	3.5	4.2	Smaller is better
Number of layers	63	51	Larger is better
Line width of metal on ceramic (μm)	90	100	Smaller is better
Separation between metal lines	450	450	Smaller is better
Propagation delay(ns/m)	7.3	9	Smaller is better
Allowed power dissipation	16.7	7.5	Larger is better
Substrate size (cm²)	160.0	112.4	Larger is better
Coefficient of thermal exp. (10-7/C)	30	35	Smaller is better
Substrate-silicon match	1.0	0.86	The closer to 1, the better
System-Level Performance	**Z1000**	**SS180**	
Module contribution to system cycle time	4.60	4	Smaller is better

The table shows a comparison of fundamental technologies and realized system performance for two mainframe processors in this empirical sample. While the Z1000 is better than the SS180 on every fundamental dimension, its system performance is worse. Using the terminology of this book, the Z1000 has higher technological potential than the SS180, but worse technological yield.

circuit design. The SS180, for example, is better at leveraging a small number of very fast connections between integrated circuits. While most of its signals may have to go a long way, its *most critical* signals do not have as far to travel as in the Z1000 design. This result stems from closer linkages between logic partitioning choices in the integrated circuits and material choices at the module level. Other advantages of the SS180 link technical choices to intimate knowledge of the production environment. One of the Z1000's outstanding technological achievements, for example, was an exact thermal match between its ceramic and metal materials. Attaining this represented a breakthrough in material science, accomplished by carefully "doping" both ceramics and metal until the desired result was achieved. The SS180 did not accomplish this technological feat, however, because it did not have to. The differences in thermal expansion between the materials could be accounted for by the way the rest of the module was manufactured. While doing so required much less work at the level of fundamental technology, it meant a much *better* understanding of how details of the manufacturing process would allow less aggressive technology choices to work as well.

The SS180 and Z1000 followed drastically different approaches to technology choice. The philosophy behind the Z1000 was to make such sophisticated (and aggressive) technical choices that the product would satisfy objectives regardless of the details of the application context. The perfect thermal match means that the materials will not crack no matter what the actual design and characteristics of the operating context are like. This proved to be an exceedingly difficult technological challenge, however, and added considerable delays and expenses to the project. By comparison, the technology choices in the SS180 used the context to their advantage and were driven by a much better understanding of that context (the system knowledge described in Chapter 2). This allowed selections to be less aggressive. Project members had deep knowledge of how details of the plant environment would interact with their technical approach, which gave them confidence that more conservative (and thus less expensive) choices could still be leveraged into a highly effective product. The higher system performance of the SS180 (and much higher R&D productivity) was therefore linked to a process that gathered and applied system knowledge to make critical technology choices.

Workstation systems and mainframe processor modules are very different types of products. Performance in the first is driven by microprocessor muscle, ASIC (application-specific integrated circuit) design, and a detailed knowledge of the computer's architecture and software environment. Performance in the second is driven largely by deep knowledge of material science and lithography and knowledge about the manufacturing environment. Both the IRIS and the SS180 achieved superior system performance in a similar fashion, however. They both used an organizational process that combined fundamental and system knowledge (product architecture or manufacturing environment) to match fundamental technologies to the details of the application context. In contrast with them, the Z1000 and the Alpha workstation had much higher technological *potential*, driven by excellence in deep, specialized expertise. Their technologies could have provided the basis for an incredibly fast product, but not all of this potential was realized. Their technological *yield* was thus considerably lower than that of their competitors, leading to a product with lower performance.

The world of skiing provides an interesting related example. Franz Klammer, an Austrian downhill champion who dominated the sport for many years, won the World Cup several times, along with several world championships and Olympic medals. His skiing style puzzled many experts, however. He was consistently clocked going more slowly than several of his more athletic competitors on any single segment of the downhill run. Yet when all the segments were integrated, Klammer consistently came out ahead. Indeed, being slower in certain parts of the downhill course actually helped him elsewhere: It enabled him to brake less around turns, losing less of the speed gained in the previous straight. Rather than optimizing speed on any given part of the race course, he would approach the entire run in a balanced fashion. In our language, while other skiers had more "potential," Klammer was consistently able to "yield" more actual performance.

Analyzing Technology Choices

The previous examples are helpful in understanding that there is more to a product than its sheer technological potential. But how do we analytically assess the achievement of a higher yield? How do we analytically connect this achievement to the approach taken? To do this we must define a methodology for measuring technological potential and technological yield.

The methodology works best when two conditions are satisfied: its product performance can be defined precisely and its upper bound can be modeled by a relatively simple algebraic expression. This is true in many cases. Many products have well-defined performance measures. Downhill skiing performance is a function of overall elapsed time; a semiconductor memory can be assessed in terms of its density and speed; disk drive performance is a function of size, weight, capacity, and access time; and engine performance can be captured by horsepower, weight, fuel consumption, and other well-defined measures. The performance of many products can also be modeled analytically (through an algebraic expression). In these cases, it is usually possible to define a model that will establish a performance upper bound (product performance in practice will always be less than what is predicted by the model). The density of a semiconductor memory is always less than the square of the width of the smallest possible circuit (the minimum line width). The time of a downhill run will always be more than the distance divided by the maximum speed.

When these two conditions are satisfied, the lowest known upper bound model is defined as the product's technological potential, given by the quantity TP. The term *potential* is justified by the fact that, given the parameters of the model, the best its performance can ever be is represented by TP. TP will be a function of a number of parameters (p_i) describing the product's characteristics, such as the properties of its materials (for example, dielectric constant and thermal conductivity), geometry (important length scales and densities, for example), or operating parameters (such as temperature and stress). TP thus represents the maximum achievable product performance given the current state of fundamental laws of physics applied to an environment characterized by the parameters p_i.[2]

By definition, a product's performance will be less than (or equal to) its technological potential. In practice, performance will be influenced by a large number of design details whose impact is not known or articulated in clearly defined models; these details range from the cleanliness of its production environment to its specific geometry. The impact of this complex set of considerations is captured in the technological yield, or TY. Because TP is defined as the lowest upper bound, TY measures how close a product actually comes to its maximum theoretical performance, and will always be less than *1* (see Figure 5-2). In other words, by definition, the details captured by the technological yield factor can only degrade the theoretical maximum performance, not improve it.[3]

$$\mathbf{P = TY*TP\ (p_1, p_2, p_3, \ldots)} \quad (1)$$

For a semiconductor integrated circuit, actual performance will be degraded by a variety of factors such as granularity, imperfections in the boundary between materials, or defects in the device geometry. These are a function of various contextual considerations, ranging from the design of the deposition equipment to the quality and temperature of the water used in cleaning the wafers. Each of these effects will decrease transistor switching speed, degrading its ideal dependence captured by analytical models. The precise extent

FIGURE 5-2

Product Performance as a Function of Fundamental Parameters(p_i)

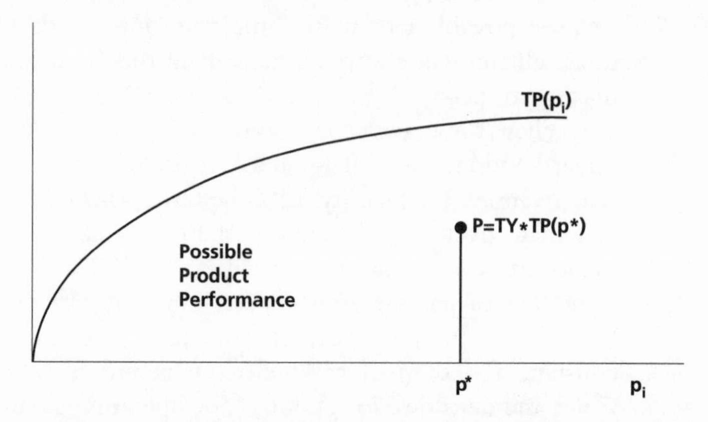

NOTE: The technological potential TP provides an upper bound given by TP(p_i). The region beneath the curve indicates the possible range in performance. The higher the technological yield (TY), the closer product performance is to the upper bound. TY is always less than 1.

of the effect, however, will be difficult to characterize *ex ante* or to articulate and capture in simple analytical models of device performance.[4]

Technological yield can be considered the R&D analog of production yield. In a realistic production environment, theoretical plant capacity cannot be completely attained, given inevitable down time, waste, and quality problems. Similarly, the theoretical potential of a technology may never be fully achieved, given the complex constraints of a real product context.

Technology Choices in Advanced Computer Processors

I now turn to the systematic application of this methodology to the study of advanced computer processors.

Empirical Approach

Mainframe and supercomputers provide a good environment for investigating technological potential and yield; product performance is well defined, can be easily modeled, and is simple to track. Although the measures and models have remained constant, the results achieved have increased dramatically over time. Major technological advances are common, but product performance measures are stable enough to allow a good analytical comparison of products and projects. The model for technological potential in this environment is simple to derive, and is shown in Appendix I.

Matching the work described in Chapter 4, the approach here focused on the processor module. The Z1000 and SS180 are typical of the empirical sample. For this analysis, I narrowed the sample to the seventeen projects that could be described by exactly the same technological potential model.[5] Doing so enabled a precise technical analysis to be combined with detailed field-based observations, project by project.

Measuring Potential, Yield, and Performance

Figure 5-3 displays the performance of the products and replicates with empirical data the conceptual sketch shown in Figure 5-2.[6] As expected, the realized performance is less than the technological potential for all products. Figure 5-4 shows explicitly the values of technological yield obtained by the seventeen product outcomes.

FIGURE 5-3

Technological Potential and Realized Performance as a Function of Gate Density

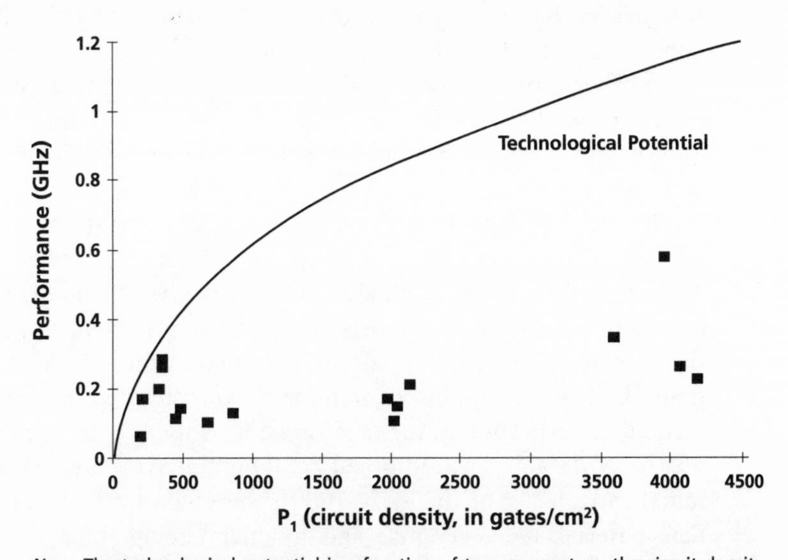

NOTE: The technological potential is a function of two parameters, the circuit density and the dielectric constant (see Appendix I). The curve is a projection of the technological potential onto the plane defined by a constant dielectric constant, ε=3.5, which was the average for the sample. The squares represent the realized performance for the seventeen projects.

FIGURE 5-4

Technological Yield Values for the Products in the Empirical Sample

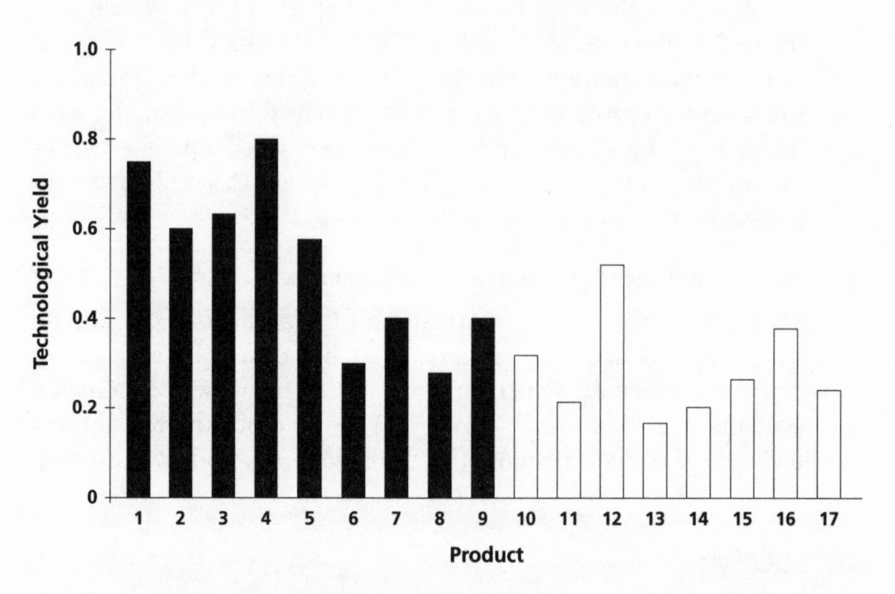

NOTE: The figure displays the technological yield for all seventeen products in the sample. Products in the first generation are indicated by black bars; products in the second, by white bars.

It indicates that some projects come quite close to realizing the full potential of their technology, obtaining yield values close to one, while others only realize a small fraction of the potential. Such differences can be a clear source of competitive advantage in this industry, since computer performance is a major driver of market success.

Figure 5-4 includes data on two product generations, the first comprising products introduced before January 1, 1990, and the second products introduced after that date. The second generation generally exhibits lower values of technological yield, suggesting that it may be less technologically mature and thus harder to integrate than the first. This assumption is consistent with the path of technology evolution in the environment; while the first generation was based largely on traditional ceramic materials (primarily Alumina) introduced in the early 1960s, the second relied heavily on new materials (glass ceramics and polymers) being employed in this type of product for the first time.

Table 5-3 investigates the relationships between product generations and shows that the second has higher technological potential,

TABLE 5-3

Differences in Product Outcome Between Generations

	Mean and Standard Deviation for Generation 2	Mean and Standard Deviation for Generation 1	t value	p (1-tail)
Performance (GHz)	0.24 (0.15)	0.16 (0.07)	1.53	0.073
Technological potential (GHz)	0.88 (0.29)	0.31 (0.09)	5.61	0.0001
Technological yield	0.27 (0.11)	0.52 (0.19)	–3.20	0.003

Values in parentheses are standard deviations. The total sample contained seventeen points, of which nine were in the first generation (introduced before January 1, 1990) and eight in the second.

higher realized potential, and lower technological yield than the first. The difference in realized performance is only significant at the 10 percent level, however, because of the large variation between projects in the second generation. In moving from one generation to the next, firms invested in fundamental technologies, increasing the technological potential of their products. Although doing so increased, on average, product performance, it also increased the challenge inherent in the development process, indicated by an increased variation in realized performance and low technological yield.

These results indicate that while high technological potential is necessary, it is not sufficient to generate a product that leads in its generation. Achieving high technological yield is critical in differentiating the highest performance products.

Project Performance and Product Outcome

How could different competitors obtain such different values of product performance, technological potential, and yield? The most obvious explanation is that higher *product* performance was obtained at the expense of *project* performance. In other words, by taking longer and using more people, a better yield and potential will be obtained. Table 5-4 shows that this is not the case, however.

The table investigates the relationship between project resources (given by person-years, as in Chapter 3), lead time, product performance, technological potential, and technological yield.

TABLE 5-4

Correlation Coefficients Among Technological Yield, Technological Potential, Product Performance, Project Resources, and Total Project Lead Time

	Technological Yield	Technological Potential (GHz)	Product Performance (GHz)	Project Resources (person-years)
Technological yield	1			
Technological potential (GHz)	–0.527*	1		
Product performance (GHz)	0.257	0.614**	1	
Project resources (person-years)	–0.495*	0.365	–0.186	1
Total lead time (years)	–0.172	–0.152	–0.465*	0.540*

A * indicates significance at the 5 percent level, and ** indicates significance at the 1 percent level.

It shows that better outputs are not correlated with greater resources; there is no statistically significant association between project resources and either realized performance or technological potential. Moreover, technological yield correlates *negatively* with project resources: On average, more efficient projects achieved product outcomes with better yield. This puzzle is not resolved by trivial factors, such as increased product cost or increased development speed. First, total lead time (measuring the time elapsed between the earliest targeted research and market introduction) also correlates positively with project resources, indicating that the more efficient projects were also the fastest. Second, there is evidence that the level of project resources is not correlated with high product cost.[7]

This surprising finding has a deeper implication, which supports the conjecture at the beginning of the chapter. Achieving a better match between technology and context not only helps product performance, but also improves project efficiency and speed. Stated differently, this study has formally shown that good projects (the high performers, in the terminology of Chapter 4) come up with systematically different products, achieving a better match between fundamental technology and system context.

How is a good match between technology and system achieved?

The next section focuses on this problem by exploring the relationship among the characteristics of organizational process, technological potential, and yield.

Relating Product to Process Characteristics

The data shows that both technological potential and yield values can vary substantially among competitors—and this variation should reflect differences in process. Technological potential should be associated with the generation and accumulation of domain-specific knowledge. Advantage should be achieved through excellence in research in clearly articulated scientific and technical domains. For its part, technological yield should instead measure a very different type of contribution. It should be linked to the interactions between those domains and all other factors relating to the context of the product, captured by the base of system knowledge mentioned earlier—including details of the production process, user environment, and product architecture.

This is consistent with the evidence provided in Table 5-5, which displays correlation coefficients between elements of the process followed in a project, and the resulting technological potential and yield. Technological potential is associated with a tradition of research in the fundamental knowledge domains defined by the potential model. Technological yield is associated with the existence of a targeted technology integration process, conducted by a group staffed with dedicated individuals. (The methodology for assessing organizational process variables was described in Chapters 3 and 4.)

TABLE 5-5

Statistical Correlations Between the Basic R&D Process and Product Characteristics

Characteristics of the R&D Process	Technological Yield	Technological Potential
Tradition of research in fundamental knowledge domains defined by the model	0.202	0.661**
Targeted integration process, with dedicated group of individuals	0.753**	0.178
Development team dedicated to the project	0.173	0.242

The double asterisk indicates significance at the 1 percent level.

Interestingly, neither technological yield nor potential is associated with a dedicated development team, one of the leading measures of organizational integration in the development phase (see, for example, Wheelwright and Clark, 1992). This result suggests that product performance is more closely related to the early project stages of research and technology integration than with the later stages of detailed design and implementation.

While these results show that research is a primary driver of technological potential and that technology integration is a primary driver of technological yield, the relation between these variables is complex, and needs to be studied more thoroughly. The following sections present additional evidence. The analysis is broken down by stage; it starts with the research stage, continues with technology integration, and finishes by looking at development.

Research Process

The relationship between research process and product performance is shown in Table 5-6, which highlights interesting differences between the impact of research on technological yield and technological potential. First, as noted, the contribution of technological potential to product performance is positively correlated with access to a deep knowledge base in the domains defined by the technological potential model. This is indicated by the existence of a dedicated research group and, especially, by a tradition of research aimed at improving the fundamental parameters identified by our model. A high contribution is not correlated with the external acquisition of technology, either from suppliers or from other business groups in the company (the correlation is very close to zero). Such a picture is consistent with the findings of Cohen and Levinthal (1990) because it indicates that even when technologies are developed externally, they still require a critical mass of internal knowledge for them to be absorbed. Finally, the achievement of high technological potential was not correlated with intense communication within the product development organization, with experience in product development, with experience in the previous application of the specific technology. The measure of technological potential, therefore, reflects disciplinary depth in research, not the extent to which research investigations may be influenced by experience with the application context. This is consistent with our expectations.

Table 5-6 also shows correlations between research process and technological yield. The correlation with technology acquisition is significant and positive; so is the correlation with the existence of

TABLE 5-6

Correlations Between Observed Characteristics of the Research Process and Values of Technological Potential and Technological Yield Achieved

Characteristics of the Research Process	Technological Yield	Technological Potential
Tradition of research in fundamental knowledge domains defined by the model	0.202	0.661**
Dedicated research group exists	0.476*	0.502*
Chosen technology from suppliers	0.751**	0.0358
Chosen technology from other group	0.717**	0.179
High research to development communication	0.338	–0.00775
Researchers introduced a product before	0.358	0.270
Technology introduced in a product before	0.403	–0.130

All coefficients are normalized by dividing the values of yield of technological potential by the highest achieved in a given generation. A * indicates significance at the 5 percent level, and ** indicates significance at the 1 percent level.

an internal research group, chosen technology from suppliers, and chosen technology from other groups. This result suggests that high technological yield is linked to the *breadth* of technical options. The more groups involved in the research process, the broader the options presented for later integration, and the higher the likelihood of crafting a technical concept that matches fundamental technology to the application context. The impact of communication and experience in research is positive, but not significant at the 5 percent level.

Technology Integration Process

As expected, many elements of the technology integration process have a significant correlation with technological yield. The results shown in Table 5-7 are strikingly sharp. While almost every aspect of technology integration is correlated with yield, only one aspect is correlated with potential.

The existence and dedication of a group focused on technology integration tasks correlate significantly with technological yield, as does the direct contact of members of the group with aspects of the environment in which the technology must function, such as plant,

TABLE 5-7

Correlations Between Observed Characteristics of the Technology Integration Process and Values of Technological Potential and Yield Achieved

Characteristics of the Integration Process	Normalized Technological Yield	Normalized Technological Potential
Integration group (IG) exists	0.594*	–0.0453
Core of IG is dedicated	0.753**	0.178
IG in day-to-day contact with plant	0.747**	0.144
IG fixes production problems	0.684**	0.0497
Manufacturing engineering group not on critical path	0.684**	0.0497
IG relocates at uncertainty source	0.610*	0.186
IG drives production equipment choice	0.527*	0.286
IG interacts with system group	0.596	–0.100
IG interacts with component groups	0.301	0.241
Technical expert manages IG	0.646**	–0.0787
IG members have T specialization	0.750**	0.228
Firm introduced similar product before	0.259	0.381
IG introduced similar product before	0.623**	0.308
Technology integration index	0.593**	0.033
Experimental capacity	0.162	0.505*
Experimental similarity	0.699**	0.194

All coefficients for potential and yield are normalized by dividing the values of yield of technological potential by the highest achieved in a given generation. Two variables, experimental capacity and representativeness, were added after the semiconductor study was started and were not originally included in the technology integration index. A * indicates significance at the 5 percent level, and ** indicates significance at the 1 percent level.

production process, equipment choice, and other sources of contextual uncertainty. Technological yield is thus positively correlated with the existence of a group that spans the boundary between the technology and the context in which it is to be introduced, generating and applying system knowledge.

Also interesting is the lack of significant correlation between most coded characteristics of the integration process and the technological potential. This finding is consistent with a model wherein the fundamental technical breakthroughs are not driven by the integration group, but by the research work that precedes it. And while the correlations are not usually significantly positive, they also are not significantly negative. In other words, a strong technology integration process does not hurt the development of an aggressive, novel technology.

In one case the impact of technology integration on technological potential is significant and positive: the existence of infrastructure for extensive experimentation. Achieving high technological potential is aided by running many experiments—the more aggressive the chosen technology, the more the number of experimental trials necessary to make sure that a satisfactory solution is found. In the absence of this capability, the fundamental technology might never reach the marketplace.[8] This point will be elaborated in Chapter 7.

Development Process

Finally, Table 5-8 investigates the development stage. Variables that are traditionally the most critical in implementing a project do not seem to have a great impact on either technological potential or yield. Particularly striking is the lack of association with the dedication of the project team and with the existence of a project manager, factors which show up as important in studies aimed at the management of design projects (see, for example, Wheelwright and Clark, 1992; and Bowen, et al., 1994). This finding suggests that once a concept is selected, project team integration only has a limited impact on product performance.

Some aspects of development do have an impact on potential and yield, however. The responsibility and influence of the project manager, for example, has a significant association with the variables. Particularly interesting is the strong association between the influence outside engineering and the technological potential. It is not clear which way the relationship goes, though. Strong project management may increase the aggressiveness of a technology. Alternatively, if the technology in a project is aggressive, it is more likely that a strong project manager will be chosen.

One way to interpret the partial significance of development process variables on technological potential and yield is as follows.

TABLE 5-8

Correlations Between Observed Characteristics of the Development Process and Values of Technological Potential and Yield Achieved

Characteristics of the Development Process	Technological Yield	Technological Potential
Development team is dedicated	0.173	0.242
Project manager (PM) exists	0.337	0.00848
PM has wide responsibility	0.485*	0.121
PM influences engineering	0.337	0.00848
PM has influence outside engineering	0.202	0.661**
PM has direct contact with system group	0.539*	–0.257
Internal integration index	0.211	0.436

All yield and potential coefficients are normalized by dividing the values of yield of technological potential by the highest achieved in a given generation. A * indicates significance at the 5 percent level, and ** indicates significance at the 1 percent level.

The concept chosen during the technology integration process essentially sets a target for both technological potential and yield. Although it is difficult for the development process to improve on the target, it is still possible to fail to achieve it—through lack of leadership or cross-functional communication, for example. Project management capability would thus correlate with performance, although it would not be the primary factor in achieving it.

Technology and Its Organizational Mirror

Although many authors have argued that looking inside the "black box" of technology is important, broad-based empirical investigations linking organizational process characteristics to specific technological impact are rare. The present work attempts to fill some of this gap by using methodologies based in both the physical and social sciences—combining a technical analysis of product performance with an organizational analysis of the project. The results reveal that project performance is linked to product characteristics—faster and more efficient projects were characterized by product outcomes obtaining higher technological yield. Further analysis showed that product outcome mirrored the process that created it.

After speculating that technological potential is driven by processes aimed at gathering fundamental knowledge, I found, not surprisingly, that potential was linked to depth in specialized research. I argued that technological yield would instead be driven by processes that generate information on the match between fundamental knowledge and the implementation context. Again, in accordance with expectations, I found that the process of technology integration was associated with high technological yield.

The drivers of technological potential and yield merit further elaboration. Technological potential is not only linked to research capability. A product will have high technological potential only if it is introduced to the market rapidly, since this analysis compares its fundamental technologies to those of other products on the market at the same time.[9] If an organization is extremely ineffective in all stages after research, the product will be introduced with substantial delays, making its potential at the time of introduction low. As such, other factors, such as experimentation capacity and project management influence, make a difference.

Technological yield is driven by very different factors. Though it is not the result of excellence in fundamental research, it is also not simply the result of good design or simultaneous engineering. Significant improvements in technological yield were consistently linked to decisions made in the early project stages, going well beyond traditional design tasks. Once the fundamental technological path was defined, opportunities for the broad, architectural solutions that characterized significant technological yield improvement rapidly disappeared. The achievement of high yield frequently involved the execution of "gap-filling science" (Allen, 1977) as well as fundamental changes in product architecture and production equipment.

More than simply clever design, high technological yield is founded on a particular orientation in the technology integration process. It is linked to an early focus on systemic interactions between knowledge bases and a tendency to explore how the system as a whole will behave before making commitments to individual fundamental choices—what I've called system-focused technology development (see Iansiti, 1995a-c). The essence of the approach was captured by a U.S. semiconductor engineer taking part in a joint venture between his company and a Japanese partner:

> What they [researchers in the Japanese company] do never ceases to amaze us. They seem to procrastinate on the most fundamental

> choices until the last possible moment, at which point they have a great idea of how the whole system fits together. Even though we are much more integrated than we used to be, our orientation is still to spend a long time figuring out the basic pieces and do the system integration later. We get a lot of early input from manufacturing as to what might work, but we don't actually put the system together until later. Somehow, they do the system integration first, almost in the research stage, and finalize the components later. As a result, they come up with a better overall solution. I think some day we will get there too, but we certainly aren't there now.[10]

In conclusion, the contributions of this chapter are both methodological and conceptual. The methodology is general and can be extended to a variety of products, as long as they have clearly defined performance or quality measures; it enables a quantitative analysis of the relationship between technical outcome and organizational process in a wide variety of industrial environments. From a conceptual perspective, the work shows that different technology integration processes are indeed reflected in systematic differences in product outcome. It demonstrates that the processes driving the generation of integrative system knowledge are truly different from those driving the generation of fundamental technological knowledge and that each has a distinguishable impact on product performance.

Notes

1. The names of the two products are disguised, but their characteristics correspond to those of actual products included in the sample.
2. Typically, technological potential models are widely disseminated, a result of the widespread acceptance of the scientific concepts underlying most of the products we use. In the semiconductor industry, for example, analytical models have their roots in theories (electromagnetism, basic quantum mechanics) already well established by the 1930s. Models for estimating increases in technological potential are therefore not usually considered proprietary and are well disseminated throughout the industry, published by national and international research and standard-setting organizations such as Semetech or the National Advisory Committee on Semiconductors.
3. Since TP is defined as the lowest upper bound, existing analytical models may be "better" predictors of product performance. Calling the best predictor model M, this methodology assumes that the difference between M and TP is small. This assumption is not essential, but is motivated by clarity and simplicity. This methodology could be based on M rather than TP, and the only difference

would be that the yield factor could then be greater than one. However, the best predictor model frequently is a lowest upper bound (that is, M = TP). This is because fundamental analytical models that predict product performance are generally based on a number of assumptions describing ideal conditions, such as material uniformity or thermodynamic equilibrium. In practice, the details of a real environment tend to make performance worse than theoretical predictions. It can be shown rigorously, for example, that the efficiency of an engine is always less than one, which is the result of the third law of thermodynamics.

4. Among many other factors, the impact of granularity on performance will depend on the properties of the boundaries between the grains; these are in turn a function of myriad considerations, such as the cleanliness of the deposition environment (for example, the existence of minute concentrations of grease in the deposition chamber), the local temperature of the material at the time of grain growth (which may be affected by the way the wafer is attached to the deposition chamber, or the nature of the cooling system), as well as a large number of unknowns. These considerations are much too complex and ill defined to be captured precisely in simple analytical models of device performance.
5. The remaining ten projects were based on CMOS (instead of bipolar) integrated circuits and were aimed at a slightly lower segment of the high-performance computer market. They were excluded from the analysis because their performance vector was different from that of the seventeen bipolar projects.
6. The technological potential curve—$TP(\varepsilon,g)$—is plotted as a function of one of the two parameters, gate density. The second parameter, the dielectric constant, is held fixed at *3.5*, which is less than or equal to the actual value for all products plotted. Also shown is the realized system performance *P* for all products in the sample. Values of t_d were drawn from either technical publications or interviews with project members. The parameters *g* and ε were found by looking at published product characteristics and were confirmed through interviews. (See Appendix I for a detailed discussion of the technical analysis.)
7. While precise data on production yield could not be gathered for all seventeen competitors, the most productive projects were not associated with particularly low yield or high variable cost. In any case, the fundamental cost driver in mainframe and supercomputer hardware is development cost, not variable cost. This is especially true for processor interconnect substrates, whose variable cost is a small fraction of the cost of the entire system (typically <0.1 percent). Their investment cost, which correlates highly with R&D productivity, is, instead, of the order of one fourth of total investment.
8. It is important to remember that the definitions of both technological potential and yield are based on products introduced to the market and are compared to products in the same generation. Both measures are therefore *ex post* evaluations, not measures of intent. If a project suffers substantial delays, the value of technological potential will decrease. Chapter 7 takes more careful account of the time dependence of these variables for the semiconductor data.
9. In this chapter, time is defined by product generation only for the mainframe data. A more precise treatment is given in Chapter 7 for the semiconductor data.
10. Interview conducted by author during field research from 1990 through 1995.

Problem-Solving

chapter 6

Foundations

PROBLEM SOLVING IS the engine of knowledge creation. Whether the problem is reducing the time needed to commute to work or improving the yield in a production process, the path taken to approach a solution will lead to the search for new knowledge, to its interpretation, and to its application and internalization in creating and validating a solution. In fact, a number of academic writers have described knowledge building as a sequence of problem-solving efforts (Von Hippel, 1994; Dosi and Marengo, 1993; and Iansiti and Clark, 1994). Problem-solving efforts therefore provide a focused lens with which to examine the foundations of the organization's innovation process.

The previous chapters showed how the process of technology integration is critical to performance in complex and novel empirical settings. It is associated with *project* performance, indicated by development lead time and productivity, and it is critical to *product* performance, being associated with the match between fundamental technology and system context. This chapter examines the microscopic mechanisms through which these associations actually occur. To do this, it zeros in on detailed observations of problem-solving efforts, and shows that problem-solving behavior that navigates across diverse knowledge domains is at the heart of technology integration, and is critical to both project and product performance.

The analysis also reveals great consistency in the empirical results; the three different methodological approaches discussed here and in Chapters 4 and 5 produce a highly consistent set of findings, each emphasizing the integration of knowledge. A proactive process aimed at integrating knowledge of technology and context leads not only to high project performance (as described in Chapter 4), but also to "more integrated" outcomes, characterized by higher technological yield (as described in Chapter 5). This in turn correlates with the integrative breadth of problem-solving efforts observed at the microscopic level, as described in this chapter.

Beyond analytical comparisons, the problem-solving field work also generated rich, detailed accounts of how technology integration occurs in practice. In some instances I found different organizations tackling exactly the same problem. These examples are extremely useful in understanding the linkage between technology integration and performance, by contrasting the approaches taken and the ensuing results. I start this chapter by describing some of these accounts to set the stage for my analysis and discussion.

Problem Solving in Mainframe Processors

The Buckling Problem

The research on problem solving in computer processor modules revealed several examples of different organizations attempting to resolve similar technical problems. Such is the problem of ceramic buckling, understanding of which requires a brief introduction to ceramic substrate processing. A substrate module is made up of many thin layers of a ceramic material. The layers are coated with metal to create a pattern of connections. The layers are then stacked one on top of the other, and sintered together at high temperature. As a result, the substrate coalesces as a single, reliable unit, while shrinking by a substantial fraction (usually about 20 to 30 percent).

In the development of a high-performance substrate, a critical and common problem is avoiding substrate buckling. Substrates buckle when they do not shrink uniformly; if the edges shrink more rapidly than the center, the result looks more like an ashtray than a flat board. This is a major problem, since the integrated circuits to be mounted on top of the substrate require a flat surface.

Two organizations faced the same buckling problem in simultaneous projects conducted during the late 1980s. Both were developing similar modules, with similar performance requirements and made with similar materials. In both cases, the organizations detected the buckling problem during the early stages of technology integration, long before the module was to be transferred to the plant.

When the problem was detected in the first organization, a task force of material scientists was pulled together to solve it. Team members included world-renowned specialists with profound knowledge of the requisite fundamental knowledge base (the physics of glass ceramic materials). The team worked on the composition of the ceramics, experimenting with the addition of small amounts of "dopant" materials. These additions influenced the shrinkage rate of the ceramic, with the goal of creating a uniform rate across the entire substrate. After many months of effort, working intensely in the small-scale sintering ovens in the research laboratory, the team was successful and virtually eliminated the buckling problem.

Unfortunately, when the sample was transferred to the pilot plant, populated by larger ovens, the buckling problem reappeared. Once again, the problem was resolved by the same team of scientists, who came back and readjusted the composition of the materials and the temperature profile of the ovens. This caused a project delay of several months. Moreover, when the substrate was transferred from the pilot plant to the manufacturing facility, the problem appeared yet again. The higher production volumes required even larger ovens and a faster production process, which again changed the dynamics of the sintering process. Once again, the team of scientists was assembled, and once again, after several months, the problem was finally solved. Over all, this problem caused a project delay of about a year and a half and cost many person-years of added project resources.

The second organization facing the same buckling problem approached it in a fundamentally different way. Problem resolution was not entrusted to a specialist team but to the individuals in the technology integration group, who had responsibility for specifying technology choices and architecture for the entire module. Group members had diverse backgrounds, detailed knowledge of the application context (primarily the production facility), and direct access to the plant for experimentation. Their first step was to prototype the module production process in the actual production plant. They were thus able to observe the buckling problem in the actual context and characterize its precise extent.

After extensive discussions and brainstorming, the group decided that the extent of the buckling could actually be acceptable in production, as long as they found a different way to create a flat surface on the substrate. Some team members suggested planarizing (that is, creating a flat surface by coating) the substrate with a polymer. This added a production step but eliminated the problem completely. It never returned. The solution enabled this company to introduce its product more than a year ahead of the other firm. The savings in engineering resources more than made up for the added expense of the additional production step. Moreover, the one-year timing advantage yielded very significant profits and market share increases.

The Foundations of Technology Integration

The buckling example highlights the microscopic roots of effective technology integration. As argued in Chapter 2, decisions need not appear momentous to profoundly affect a project's performance. From a broad perspective, many firms in the mainframe study made similar decisions at the strategic level. They had similar alliances, chose similar architectures, researched similar fundamental technologies, and emphasized similar technical disciplines. At more microscopic levels of analysis, however, organizations differed dramatically. Their individuals had different skill sets, solved problems in profoundly different ways, and ran projects in ways that emphasized widely different approaches and methodologies. These differences appeared to have an enormous impact on product and project performance. They could cause a large variation in project efficiency and speed—in some cases even making the difference between profitability and industry exit. Thus, in the buckling case, differences in the approach taken to solve a single technical problem were responsible for large differences in market share and profitability.

That example also shows the power of merging fundamental knowledge with knowledge specific to the application context. The time the second organization spent experimenting with the new module in the plant was invaluable. Information learned from the prototypes was used to change the basic architecture of the module, using an entirely different technology (the polymer layer) to effectively bypass the challenging buckling issue. Finally, the example illustrates the tight relationship between problem-solving behavior and organizational process. Why didn't the first organization follow the same path? Part of the answer lies in the individual skill set of the problem-solvers. Also important, however, are characteristics of its organizational processes, which influence a wide variety of factors from the access to experimental facilities to the flexibility to change module architecture to the make-up and influence of the technology integration team. The more efficient solution was not merely born of cleverness or creativity but also of the right process and infrastructure. This provided the conditions for a broad and flexible approach to problem solving.

Problem Solving in Computer Server Development

The mainframe example above is complemented by a look at the study of server development. The following account is drawn from the later stages of the Lego project completed by Silicon Graphics in the second half of 1996 and described in detail in Chapter 7. The information is disguised to protect company confidentiality.

This problem-solving account focuses on the bring-up stage of the project, immediately following power-on, the moment of truth when simulation models are finally translated into hardware and the computer is first turned on. When this happens, a computer will typically not function—and even if it does, it typically crashes after a few seconds. Many problems must be solved before the computer system works reliably and is ready for shipment to a customer site. The bring-up stage is a rapid cluster of problem-solving iterations aimed at resolving all the issues that the simulation models have not picked up: the timing of signals, bugs introduced by the integrated circuit vendors, signal integrity problems, to name a few.

The Lego system was turned on in March 1996, when Silicon Graphics engineers received their first four circuit board sets from the vendors. The four systems were immediately set up next to each other in the lab in Building 7 of SGI's headquarters in Mountainview, California. The effort to debug the boards was run as a three-shift operation, to maintain a constant effort on problem-solving activities, twenty-four hours a day. At any time, about a dozen engineers were in the lab. Each system was attended exclusively by one of four groups of two engineers (one hardware and one software expert each). The others present included more experienced people who would observe, offer advice, and make sure that problem-solving efforts happened in a coordinated fashion.

Look at a typical problem encountered in this process. One group of engineers discovers that their system "hangs" (stops working) when it runs a certain instruction sequence. They start problem-solving activities by looking at the system through a logic timing analyzer. This equipment helps them replicate the problem many times while observing how different parts of the computer operate. After a few hours of work (and many thousands of simulated computer crashes), the two engineers discover that their machine is hanging because of an unexpected delay in a section of the new ASIC one of them had designed. They examine the circuit schematic and finally find the root cause of the timing problem: part of the physical path involved in executing one of the instructions is longer than it should be, which causes the instructions to be completed out of sequence, which in turn makes the computer crash.[1]

The engineers discuss two basic paths to solving the problem (with several other experts, who by now are hovering, lending advice). The first is redesigning the routing on the problem chip. This entails calling the vendor and sending the specification tapes back with a new routing designed by hand, optimized so that the instructions are completed in the appropriate sequence. However, the vendor will take two weeks to fix the chip (this is considered a very fast turnaround) and charge about $100,000 for the changes. Although this eliminates the root cause of the problem, the cost and delays are prohibitive. The second path involves implementing a software "workaround," as suggested by a software engineer. This workaround changes the operating system software that the engineer helped design, so that the computer will still work with the flawed integrated circuit. The change enables the code to flag the problematic instruction; every time it comes up, the operating system inserts a "wait cycle" in the instruction sequence, thereby ensuring that the instructions are still completed in the right sequence, chip flaw notwithstanding.

By now the discussion involves additional people, including the project leader and several key architects. Several other approaches are suggested, ranging from implementing circuit board changes to other software tricks. The workaround begins to gain support. The impact on performance would be minimal, since the wait cycle would only be used on average every ten thousand instruction cycles or so. Moreover, it would only take a few hours to make the change; the cost of the change, at face value, is therefore almost nothing. After some additional discussion, the group decides to go with the workaround. After five hours of intense effort, the problem is solved.

The two approaches to problem solving exhibited in this

instance are similar to the two approaches followed in the mainframe example. The first is narrower, targeting the specific domain (chip routing in this case) that caused the problem in the first place. The effort would be limited to that domain until the problem is solved. The second approach involves a broader problem-solving path. It merges knowledge from two diverse domains (ASIC design and operating-system development) to create the problem solution.

Figure 6-1 shows some estimates internal to Silicon Graphics. When we asked how often software workarounds were used to fix ASIC problems, experienced SGI engineers estimated that it was about 70 percent of the time. Ten percent of the problems were not serious enough to eliminate, and could be incorporated into the workstation's design. An additional 10 percent required a software workaround in combination with a minor change in hardware at the circuit board level. An additional 5 percent involved a change at the circuit board level. The engineers estimated that only 5 percent of the time was the root cause of the problem—the ASIC design—actually changed.

FIGURE 6-1

Approaches to Solving ASIC Hardware Problems in Workstation Development

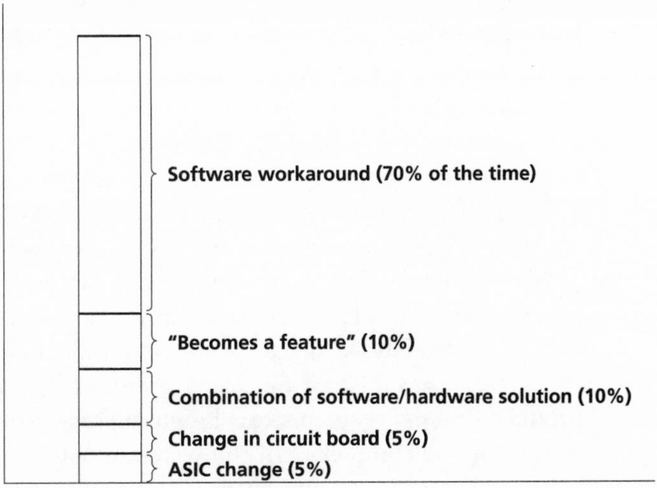

Software workaround (70% of the time)
"Becomes a feature" (10%)
Combination of software/hardware solution (10%)
Change in circuit board (5%)
ASIC change (5%)

NOTE: The data provides a rough estimate only.

SOURCE: Internal company estimate.

This works because the server (or the mainframe processor) is a complex system, characterized by many knowledge domains; hence, there will usually be many paths available to solving a given problem. An effective organization, approaching problem solving from a truly integrative perspective, will not constrain itself to a narrow discipline. It will balance the many possibilities available and arrive at the simplest way to implement a robust and effective solution to the problem. In workstation design, software is frequently simpler to change than hardware and it is thus picked as the favored path. This does not mean that the solution is less effective or robust. It may even be more robust than the more obvious solution, since the additional wait cycle may make other instructions work more reliably as well. Redesigning the routing on the ASIC might instead cause problems with other instructions. In the mainframe example, the broader solution was clearly also the more robust one.

Critical Themes

The above examples are typical of many found across the diverse empirical environments studied, and they raise four broad themes that are particularly critical to problem solving in novel and complex environments.

First, the most effective solution to a given problem may not involve eliminating its most obvious cause. Thus, the software workaround may perhaps be less satisfying than redesigning the ASIC, but it is much faster and more efficient, perhaps even more robust, and has minimal product performance cost. So while it is important to find the root cause of a problem in order to determine its extent, it is not necessary to fix the problem by explicitly targeting this root cause. Other, easier, and more robust paths may exist, like planarization in the mainframe example. If the system is highly complex, it is, in fact, unlikely that working on the specific cause of the problem will create the optimal solution. In the workstation example, many of the possible solutions involved software and circuit board changes. In other words, identifying the root causes can provide critical information, but analysis of the problem should not be restricted to them. By making use of the many options offered by the complex product system, problems can be eliminated in the most effective way, using the complexity of the system to advantage.

Second, problem-solving efficiency is associated with the breadth of the problem-solving approach. Product systems are complex and involve a wide variety of knowledge domains. The path of problem resolution can often benefit from sampling a broad set of these domains to find the fastest and more efficient path. The combined knowledge of SGI hardware and software engineers (or that of the entire technology integration group, in the mainframe example) enabled a much more efficient problem-solving path than the one each of them would have chosen individually. From a statistical perspective, it is highly unlikely that the most effective solution will be confined to one knowledge domain.

Third, "context-specific" breadth appears to have more impact. The software engineer in the SGI example did not help simply because she understood something about software development in general, but because she understood the details of the operating system as currently designed. This enabled the two engineers to find a solution that would solve the problem at maximum speed and minimum cost. People with knowledge of the plant context played a similar role in the mainframe example. As argued in Chapter 2, the design of a typical complex product is based largely on subtle, context-specific knowledge. This implies that context-specific knowledge should be of great value in technical problem-solving attempts.

Fourth, problem-solving behavior is linked to broader, higher level factors, such as organizational process, methodologies, and tools. The rapid and effective solution to the ASIC problem was enabled by the structure and process of the project, which made diverse expertise easily available and emphasized communication. Tools and methods were also important. Access to fast and representative experimentation (the simulation tools and logic analyzer, for example) was essential to problem-solving effectiveness, aiding in problem recognition, framing, and solution. In the mainframe example, access to the production facility played the same role.

These themes are consistent with observations from the academic literature. Many authors have studied problem solving (see, for example, Simon, 1978; Frischmuth and Allen, 1969; and Mintzberg, Raisinghani, and Theoret, 1976). Frischmuth and Allen studied problem solving in R&D and developed a model emphasizing two streams of activities: the generation of solutions and the generation of criteria (or frames of reference) for evaluating those solutions. In their classic study (1969), these authors suggested that in many situations, approaching the problem broadly, that is, "modifying the problem's dimensions," can lead to more effective solutions. They also pointed out that engineering training, which frequently emphasizes narrowly defined, "closed form" solutions to problems, may result in the failure to recognize novel possibilities.

More recently, other authors have stressed the importance of using broad frames of reference in technical problem solving, particularly in technologically novel environments. Dosi and Marengo (1993) argued that the evolution of frames of reference is the foundation for learning in an R&D organization. Schrader, Riggs, and Smith (1992) emphasized the difficulties of problem framing and solving in situations characterized by high ambiguity (also see McDonough and Barczak, 1992). They argued that the more novel a given problem situation the broader the framing process should be to reach an efficient solution. Effective problem solving in a project characterized by high novelty and complexity will therefore require broad frames of reference, emphasizing a broad search for possible solutions, reaching across multiple knowledge domains. A number of academics also underscore the value of merging context-specific knowledge with fundamental knowledge in technology development (for example, Leonard-Barton, 1988; and Tyre and Von Hippel, 1993). This academic literature is consistent with the above arguments and further motivates my observations.

The four themes will guide the subsequent empirical analysis, which follows a discussion of the methodology used.

Methodology: Problem-Solving Path Analysis

This study of problem solving is based on sixty-one case histories of problem-solving attempts encountered during mainframe processor development projects. In order to gather this data, I interviewed several individuals in each project and obtained detailed protocols of several problem-solving efforts.[2] I cross-checked the information by comparing the views of different individuals and by examining written technical accounts. The problems selected were "the most difficult technical bottlenecks" in each project, as evaluated by the senior engineering managers in each firm. The protocols were coded to construct an indicator of problem-solving breadth, which I then used investigate the themes outlined above. Table 6-1 provides a sample of the technical problems analyzed.

Testing the propositions necessitated developing a framework for analyzing the disciplinary breadth of the protocols. My approach was based on a technical analysis of the path followed in the attempt to solve the problems. The measure of problem-solving breadth is based on the number of independent knowledge domains sampled during each problem-solving attempt. I call this approach "problem solving path analysis."

FIGURE 6-2

Problem-Solving Path Analysis

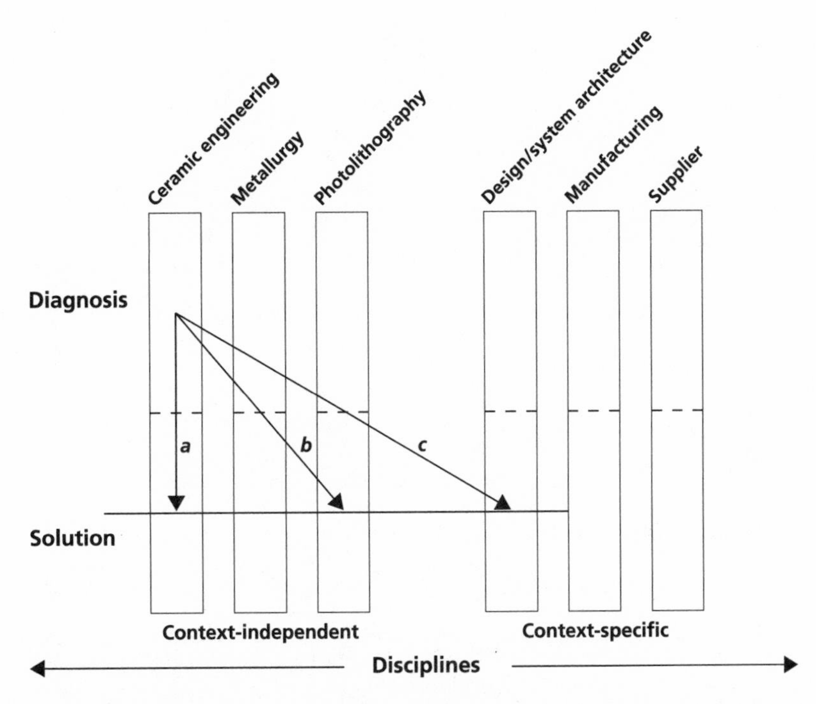

NOTE: Only a selection of the disciplines chosen is shown. Path *a* is narrow, while *b* is broad and context-independent and *c* broad and context-specific.

and *product* performance. It concludes by analyzing the association between problem-solving breadth and organizational *process*.

Problem Solving and Project Performance

Tables 6-2 and 6-3 present the results of the association between problem-solving breadth and project performance.[5] Project performance is given by person-years and development lead time, measured at the project level. The tables show that project resources are significantly associated with problem-solving breadth. The coefficients have a negative sign, indicating that the higher the breadth, the lower the person-years (and thus the higher the project's efficiency). Technical content, measured by the transistor density of the circuit (see Iansiti, 1995a-c), was included explicitly in the regressions to account for the fact that different projects had different technical challenges. Its relationship with person-years is positive,

TABLE 6-2

Regression Results on the Association among Person-Years, Problem-Solving Breadth Indexes, and Technical Content

Dependent Variable	Person-Years	Person-Years	Person-Years
Constant	364.46	294.08	110.03
Problem-solving breadth	–524.80** (205.61)		
Context-specific breadth		–478.99** (195.09)	
Context-independent breadth			63.37 (409.43)
Technical content	576.28*** (183.94)	549.01*** (186.22)	589.21*** (209.46)
F-test	8.46	8.13	4.07
Adjusted R-squared	0.374	0.363	0.197
Residual degrees of freedom	23	23	23

The table displays regression coefficients unless otherwise indicated; numbers in parentheses are standard errors. A * indicates significance at the 10 percent level, ** significance at the 5 percent level, and *** significance at the 1 percent level.

that is, the higher the technical content, the more resources necessary for the project, and significant at the 1 percent level.

Table 6-2 also displays regressions including indexes of context-specific and context-independent problem-solving breadth. The results show that while context-specific breadth is highly significant (and negatively correlated with person-years), context-independent breadth is not significantly associated with project performance. These results are similar to those found in regressions between development lead time and problem-solving breadth shown in Table 6-3. The relationship between development lead time and breadth is significant at the 1 percent level and negative, indicating that greater breadth is linked with development speed, in accordance with the first critical theme described above. The relationship between lead time and context-specific breadth is also significant and negative. The relationship between context-independent breadth and lead time, however, is not significant.[6]

TABLE 6-3

Regression Results on the Association among Development Lead Time, Problem-Solving Breadth Indexes, and Technical Content

Dependent Variable	Development Lead Time	Development Lead Time	Development Lead Time
Constant	5.73	5.46	3.98
Problem-solving breadth	–3.28*** (1.10)		
Context-specific breadth		–3.45*** (0.99)	
Context-independent breadth			1.98 (2.25)
Technical content	1.46 (0.99)	1.24 (0.94)	1.45 (1.15)
F-test	5.67	7.51	1.32
Adjusted R-squared	0.330	0.343	0.103
Residual degrees of freedom	23	23	23

The table displays regression coefficients unless otherwise indicated; numbers in parentheses are standard errors. A * indicates significance at the 10 percent level, ** significance at the 5 percent level, and *** significance at the 1 percent level.

The difference in significance between context-specific and context-independent breadth in both person-year and lead time models is important and confirms expectations. Context-specific breadth referred to the inclusion of environment-specific knowledge domains in the problem-solving effort. These included knowledge of the firm's manufacturing capabilities, knowledge of the firm's mainframe design (outside of the module subsystem), and knowledge of supplier capabilities. In contrast, context-independent breadth included general, universally applicable, traditional disciplines, such as ceramic material science or thermal engineering. The results therefore indicate that the components of problem-solving breadth that correlated with project performance are those *specific* to the firm's environment, such as the awareness of subtle design and manufacturing considerations.

This result is consistent with the likely evolution of knowledge in an organization. Over time and as they work on different product generations, engineers are likely to become aware of the links among fundamental knowledge domains. Ceramicists are going to learn some metallurgy, since the design of ceramic multichip modules has always included layers of metallic circuits to conduct electric signals. The design of ceramic and metal layers has traditionally been closely interlinked, requiring the integration of the two knowledge bases. Over time, one would therefore expect an organization to develop a hierarchy of design decisions with branches in which metal and ceramic choices are closely related, naturally leading to problem-solving progressions that involve both knowledge bases (see Clark, 1985; and Henderson and Clark, 1990).

In contrast to well-defined, fundamental knowledge domains, context-specific knowledge is more subtle and changes more rapidly (as new manufacturing equipment is introduced or as new suppliers are selected, for instance). The interactions between context-specific knowledge and technology choices are therefore likely to experience more significant shifts as products evolve. This makes the effective application of context-specific knowledge a substantial challenge, leading to clear differences in performance between projects.

Problem Solving and Product Outcome

The previous section revealed that an approach to problem solving that combined several knowledge domains, including detailed, context-specific knowledge, was associated with the efficiency and speed of projects. In this section the performance of the product itself is the focus. The first analysis, captured in Table 6-4, shows that problem-solving breadth is associated with product performance. Like the relationship between problem solving and project performance, it is context-specific breadth that accounts for the significant association.

We can go further. Chapter 5 introduced a methodology to differentiate between two drivers of product performance: technological potential and yield. The first should be linked to the depth of fundamental domains; the second should be linked to the effectiveness of the processes, matching fundamental knowledge with knowledge of the application context. Since the measure of context-specific breadth is a direct indicator of the latter, it should be associated with technological yield, but not with technological potential. Tables 6-5 and 6-6 show that this is indeed the case. Table 6-5 displays a series of simple regressions between technological potential and problem-solving breadth. Technological potential is not

TABLE 6-4

Regression Results on the Association Between Product Performance and Problem-Solving Breadth

Dependent Variable	Product Performance	Product Performance	Product Performance
Constant	0.095	0.126	0.226
Problem-solving breadth	0.232** (0.078)		
Context-specific breadth		0.265*** (0.065)	
Context-independent breadth			–0.161 (0.156)
F-test	8.80	16.82	1.06
Adjusted R-squared	0.328	0.497	0.004

The table displays regression coefficients unless otherwise indicated; numbers in parentheses are standard errors. A * indicates significance at the 10 percent level, ** significance at the 5 percent level, and *** significance at the 1 percent level.

correlated with any indicator of breadth. The association is not significant and neither negative nor positive. Table 6-6 displays associations between technological yield and breadth. The correlation between yield and breadth is significant, particularly with the context-specific indicator.

These results show that different problem-solving behaviors lead to systematically different project outcomes. I compared two independent data sets, one on product performance and one on problem-solving process. The two were based on methodologies developed independently, with data gathered two years apart from each other. The association between technological yield and context-specific problem-solving breadth is significant at the 1 percent level. These results therefore are a robust confirmation of the argument made in Chapter 5: Achieving technological yield is linked to behavior that merges domain-specific knowledge with knowledge of the specific context of application.

Problem-Solving Approach and Organizational Process

The premise of this chapter is that problem solving can serve as a lens on the firm's broader organizational processes. This is tested here by looking at the association between organizational process

TABLE 6-5

Regression Results on the Association Between Technological Potential and Problem-Solving Breadth

Dependent Variable	Technological Potential	Technological Potential	Technological Potential
Constant	0.506	0.554	0.226
Problem-solving breadth	0.168 (0.297)		
Context-specific breadth		0.101 (0.284)	
Context-independent breadth			0.145 (0.490)
F-test	0.319	0.125	0.088
Adjusted R-squared	—	—	—

The table displays regression coefficients unless otherwise indicated; numbers in parentheses are standard errors. A * indicates significance at the 10 percent level, ** significance at the 5 percent level, and *** significance at the 1 percent level.

and context-specific problem-solving breadth, looking specifically at the organizational process for technology integration and measuring it as in Chapters 4 and 5.

Table 6-7 displays the results of three regression models that relate technology integration process to problem-solving path. The problem-solving breadth index is significantly associated with the technology integration index. Similarly, context-specific breadth and the technology integration index are significantly related. In contrast, context-independent breadth is not associated with the technology integration index. This shows that project-level characteristics of the technology integration process indeed correlate with behavior at the problem-solving level. As expected, a focused process for technology integration, emphasizing experimentation and the accumulation of experience, is associated with behavior that merges fundamental, disciplinary knowledge with detailed knowledge of the application context.

Table 6-8 describes additional findings. It shows some important characteristics of the problem-solving attempts analyzed, grouped by type of organization. The distinction between system-focused organizations and other organizations was based on the

TABLE 6-6

Regression Results on the Association Between Technological Yield and Problem-Solving Breadth

Dependent Variable	Technological Yield	Technological Yield	Technological Yield
Constant	0.260	0.285	0.473
Problem-solving breadth	0.317** (0.149)		
Context-specific breadth		0.428*** (0.119)	
Context-independent breadth			–.415 (0.258)
F-test	4.54**	12.94***	2.59
Adjusted R-squared	0.181	0.428	0.090

The table displays regression coefficients unless otherwise indicated; numbers in parentheses are standard errors. A * indicates significance at the 10 percent level, ** significance at the 5 percent level, and *** significance at the 1 percent level.

TABLE 6-7

Regression Results on the Association Between Problem-Solving Breadth Indexes and System Focus

Dependent Variable	Problem-Solving Breadth	Context-Specific Breadth	Context-Independent Breadth
Constant	–0.0747	–0.218	0.145
Technology integration index	0.0615*** (0.0082)	0.0636*** (0.0094)	–0.0024 (0.0086)
F-test	55.93	45.82	0.084
Adjusted R-squared	0.687	0.656	–0.038
Residual degrees of freedom	24	24	24

The table displays regression coefficients unless otherwise indicated; numbers in parentheses are standard errors. A * indicates significance at the 10 percent level, ** significance at the 5 percent level, and *** significance at the 1 percent level.

TABLE 6-8

Problem Characteristics and Outcomes for Different Organizational Groups

The Problem	Total Sample	System-Focused Organizations	Other Organizations
Was identified before concept selection	69%	75%	62%
Was worked on before concept selection	67%	75%	59%
Was fixed before concept selection***	44%	59%	28%
Recurred after concept selection*	25%	16%	35%
Caused delays late in the project***	56%	41%	72%

Concept selection corresponds to the end of integration activities, which occurs before detailed design has begun, as described in Chapter 2. Problems "identified" were those that project members had conjectured would be particularly challenging during the project. Problems "worked on before concept selection" were those to which the organization had devoted some resources before the integration phase was complete. Problems "fixed" before concept selection were those for which early solutions were robust enough to work in the final production environment. Problems that "recurred" were those for which early solutions did not work in the final production environment. Problems that "caused delays late in the project" were those that were not "fixed before concept selection." A *** indicates that values for the system-focused and other groups were significantly different at the 2 percent level, as indicated by a *t* test. A * indicates that the significance level was 10 percent.

technology integration index. System-focused organizations were defined as having a technology integration index greater than the sample average. The assumption was that by having a greater focus on integrating individual technologies, these organizations would gain a better perspective on the whole product and process system.

The evidence from Table 6-8 indicates that system-focused organizations approach technical problems in a significantly different manner than do the rest of the sample. The difference is not in problem *identification*, however. Many critical technical challenges in the projects were obvious enough to be targeted for early action. The real differences appear when one considers the robustness of early *solutions*. System-focused organizations were characterized by having a significantly higher proportion of problems that were fixed by early efforts and thus by having a significantly lower percentage of late problems. In other words, system-focused organizations fixed the problems in early efforts. In other organizations, individuals thought the problems had been eliminated when, in fact, they had only been postponed to later project stages.

The Roots of Technology Integration

The results above link problem-solving breadth to high product and project performance. Perhaps more importantly, they shed light on the microscopic mechanisms of an effective technology integration process. Technology integration underlies the ability of an organization to view the entire product and production system as a coherent whole, balancing the potential of individual technologies with the requirements of the context of application. The results provide evidence for how this is accomplished.

The methodology explicitly coded the path taken to solve a set of specific technical problems, documenting which knowledge domains (fundamental and context-specific) were sampled in the process. Results show that an effective technology integration process is indeed associated with behavior that integrates domain-specific knowledge with knowledge of the context of application. This grounds the conceptual argument presented in Chapter 2 and elaborated in Chapters 4 and 5 with explicit microscopic empirical evidence.

The analysis combines three data sets relating to the development of mainframe processor modules. One, on organizational process, was based on field work conducted in 1990 and 1991; the second, on problem solving, was based primarily on field work conducted in 1992; and the third, gathered in 1993, was on realized product performance, technological potential, and yield. Results reveal a great degree of consistency. Integration measured in project-level organizational process correlates with integration measured at the problem-solving level and with integration measured in the final product outcome.

The evidence suggests that, for some of the organizations, the complexity of their context is an aid in managing the novelty of their technologies, rather than a hindrance. These system-focused organizations appear to view their innovation process in a holistic fashion, using the entire product system to their advantage as they solve problems. They look for broad, lateral solutions, using knowledge from one domain to help resolve problems detected in another domain. They merge knowledge from laboratory research into a novel material with detailed knowledge of the existing production process. They combine information on new mathematical algorithms with rich feedback from customers describing potential software applications. They balance fundamental and contextual considerations in optimizing product performance. To such system-focused organizations, integration is a pervasive philosophy used to manage the novelty and complexity of introducing new products at all organizational levels.

Notes

1. The problem was caused by imperfections in the software tool that the ASIC vendor used to design the routing on the chip. This is common on complex circuits; the ASIC can contain more than a million gates, and the routing is done by a routine that sometimes fails to minimize the distance between two points.
2. This analysis was carried out for only 24 of the projects.
3. Ceramic material science has been consistently linked to ceramic substrate design, semiconductor material science to the design of silicon substrates, metallurgy to chip bonding, polymer chemistry to the application of thin-film polyimide materials and to certain chip sealing techniques, structural engineering to the design of the module housing and reliability analysis, and thermal engineering to the design of the critical cooling system.
4. An example may help to clarify the methodology. In one instance, a group of metallurgists discovered that the copper formulation they were using in their design developed troublesome holes when processed at certain temperature ranges. In response to this, they developed a novel, proprietary copper paste formulation, which resolved the problem. This problem-solving path was defined to be narrow, since only metallurgical knowledge was employed. In contrast, a group of ceramicists in a different firm found that the surface of the ceramic substrate they had developed exhibited cracks that impaired its structural stability. The resolution of this problem was not limited to ceramic engineering. The researchers found out that a polymer used in sealing the substrate cover would fill up the cracks in the ceramic, eliminating the problem. This case was classified as broad, since the identification and solution of the problem combined different knowledge bases (ceramic material science and polymer chemistry).
5. The problem-solving path analysis was not carried out for three of the organizations in the sample. The number of degrees of freedom in analysis involving problem-solving indexes is therefore lower than that for the entire sample.
6. Additionally, in contrast with the productivity regressions, the impact of technical content in lead time regressions is not significant. This may reflect the observation that organizations are frequently forced to complete the module design by a specific deadline, which is defined by the introduction of the entire mainframe system. In response, they tend to add resources to the project until they feel the deadline can be met. Productivity may therefore be a more representative performance measure for the organization. While development lead time may be driven by managerial schedule, the total person years required for project completion are a direct representation of the challenges and difficulties encountered.

Managing Technological Change

chapter 7

THE PROCESS OF technology integration helps define an effective response to technological threats and opportunities by matching the approach of a project to the firm's evolving context. The process can therefore be an important factor in the management of technological change. It can provide a critical window of opportunity for balancing the advantages of continued experience and capability with the dangers caused by organizational inertia.

Projects and problem-solving activities are not isolated events. They exist as part of the organization's environment and are a product of the organization's history. It is through their history that organizations learn, build expertise, establish routines for the management of complex tasks, and construct a foundation for the future. Over time, however, these activities are vulnerable to inertia and rigidity (see Hannan and Freeman, 1984; and Leonard-Barton, 1992), which challenge organizations during periods of technological change. Firms thus often fail when faced with technological discontinuities, overcome by competitors offering novel approaches (see Tushman and Anderson, 1986; Anderson and Tushman, 1990; Henderson and Clark, 1990; and Christensen and Rosenbloom, 1995).

Projects are not only a product of an organization's past, however, but also a critical determinant of its future (Iansiti and Clark, 1994). Organizations do not fail because of their pasts; they fail because they make inappropriate decisions, defining and managing projects in ways that do not meet the evolving needs of their marketplace. Organizational history strongly influences a project's context, but it does not by itself determine the project's success. That is defined by managerial and technical choices: the multitude of decisions linking technological opportunities to changing business requirements. These decisions, in turn, are influenced by the knowledge generation, retention, and application mechanisms described in previous chapters. Hence, the technology integration process is an important lever in defining organizational response to technological challenges.

Organizations can respond to technological challenges in different ways. Their projects can follow a revolutionary path, resulting in products that satisfy customer needs by embodying a fundamentally different technological base. Such was the case in the development of the IBM 7090, the company's first transistor-based computer, which offered its customers a large increase in functionality by employing a completely different technological foundation from its predecessor. Projects can also follow an evolutionary path. This approach leverages existing technologies, increasing their impact on performance by improving their match with their context of application. This was the case when IBM introduced the 3090 mainframe in the mid 1980s. The product achieved an increase in performance close to a factor of ten without changing its technological foundations.

Not only can both revolutionary and evolutionary paths tackle technological challenges, but both types may coexist even within the same industrial environment. In another work co-authored with Tarun Khanna (Iansiti and Khanna, 1995) evolutionary and revolutionary paths were shown to coexist in segments of the computer industry. Other authors have found similar patterns in other industries (see Utterback, 1994; and Henderson, 1995), ranging from photolithographic equipment to construction materials. The same firm might even choose different approaches for different aspects of a single new product. In the mainframe industry, we found that organizations often decided to introduce novel technologies in some subsystem, and focus on more evolutionary change in another (see Iansiti and Khanna, 1995). Projects will make different choices, selecting different technological possibilities leveraging available experience and capabilities.

Different paths to meeting technological challenges will have varying organizational implications, however (see Ettlie, Bridges, and O'Keefe, 1984; Dewar and Dutton, 1986; Eisenhardt and Tabrizi, 1995; and Tushman and O'Reilly, 1997). Revolutionary change will involve major dislocation in the relationships between domain and context-specific knowledge. Evolutionary approaches instead imply almost the opposite: the careful improvement of the match between existing fundamental technology and the context of application (see Tushman and Anderson, 1986; Henderson and Clark, 1990; and Anderson and Tushman, 1990). This means that evolutionary and revolutionary change will require different approaches to technology integration.

The analysis presented in this chapter confirms this expectation. It shows that both evolutionary and revolutionary projects can benefit from the general mechanisms of knowledge generation, retention, and application. Significant differences arise, however, in the nature of specific knowledge retention and generation processes that are correlated with performance. Effective evolution is associated with the build-up of project experience and narrowly defined, representative experimentation attempts. Revolutionary change is associated with massive, parallel experimentation combined with the capture of research experience.

Defining Evolution and Revolution

Before developing this argument, it is necessary to discuss briefly some important definitions. The analysis requires a systematic assessment of the technical path followed by a project, and the methodology introduced in Chapter 5 is useful to this effort. The measures of technological potential and yield defined there help to create good working definitions for revolutionary and evolutionary projects.

Technological potential is an indicator of technological revolution. The higher the technological potential, the more aggressive the fundamental technologies implemented, the more revolutionary the path followed.[1] A product's technological yield instead estimates the quality of the match between fundamental technology and application environment. A high yield therefore implies that technology and its context are evolving jointly. This measure sets up a good definition of an evolutionary project: the higher the yield, the closer the match between context and technology, the more evolutionary the project.

These definitions of evolution and revolution do not refer to the objectives, but to the results of a project, since this is how technological potential and yield are estimated. They are not measures of intent but outcome. It makes sense, therefore, to correlate elements of technology integration process against them to see what appears linked to successful evolution and revolution.

While somewhat unorthodox, the definitions do reflect a capability or knowledge-based perspective on technological change and on strategy (see, for example, Nelson and Winter, 1982; Tushman and Anderson, 1986; and Anderson and Tushman, 1990). A revolutionary project, as defined here, is likely to imply a discontinuous change in the capability base, since the performance increases are linked to shifts in the fundamental domains of knowledge employed. An evolutionary project is instead most likely associated with a gradual change and enhancement of the organization's capability base, since the performance increases are linked to improvements in the match between technology and application context.

Evolutionary and Revolutionary Projects in Semiconductors

Figure 7-1 shows average technological potential and yield values for a number of microprocessor companies. It is taken from the study of semiconductor projects and draws on data on all major new process introductions in the last fifteen years. As before, integrated circuit density (or gate density) is used as an indicator of performance. A positive horizontal value indicates a high revolutionary component to the project. A positive vertical value indicates a high evolutionary component to the project. The figure shows that different firms in the same environment tend to achieve very different project outcomes. Some organizations achieve high technological potential, others high technological yield, others score low on both dimensions. Interestingly, no organization in this environment obtains high average values of both.[2]

IBM scores high in technological potential and about average on yield, indicating that its products are revolutionary compared to the industry norm. Moreover, IBM's performance has improved substantially since 1992. An internal look at IBM's capabilities and processes is consistent with this analysis. The revolutionary technical approach is linked largely to IBM's research. In recent projects,

FIGURE 7-1

Average Technological Potential and Yield for Microprocessor Organizations

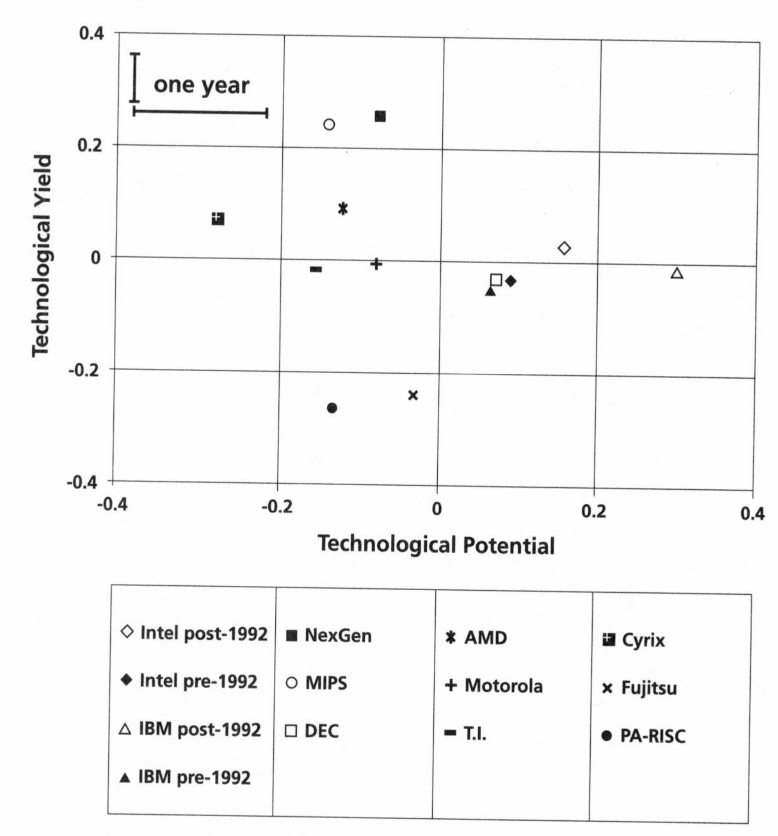

NOTE: The data plotted is the residual from regressions of the logarithm of potential or yield versus time, to take out the average trend in the industry. The logarithm of overall performance is given by adding the *x* and *y* coordinates. The bars (I, ⊢⊣) displayed in the top left-hand corner indicate how to translate the numbers in the figure into time, showing where a given organization stands relative to the industry average: for example, the plot shows that DEC is several months ahead of average in technological potential and slightly below average in yield.

for example, the high potential is caused primarily by a technology called "shallow trench isolation," which was created at IBM's T.J. Watson research laboratory in the late 1980s. This technology makes use of innovative approaches to chemical and mechanical polishing combined with reactive ion etching to planarize the surface of a semiconductor circuit layer. These allow the integrated circuit to build in height, resulting in a greater overall density.

IBM's improvement since 1992 can also be related to a substantial restructuring of its semiconductor operations, which included forming the Semiconductor Research and Development Center (SRDC) aimed at integrating new semiconductor technologies. The SRDC changed the makeup of the technology decision-making team, creating groups with combined research and implementation experience. In addition, the construction of a large new development fabrication facility significantly expanded the experimentation capacity available to the team.

MIPS, whose microprocessors are manufactured by NEC and Toshiba, is representative of a profoundly different approach. Its products, on average, score below average in technological potential but significantly above average in technological yield. This makes it weaker in fundamental technology but stronger in matching technology to its manufacturing context. The yield is linked to more efficient routing that increases the average circuit density without the smaller line widths and increased layers of its competitors. This can be traced back to the clever design created by MIPS engineers and to the traditions of manufacturing capability at both NEC and Toshiba.

A Model for Managing Revolution and Evolution

This section provides a more systematic analysis of the relationship between the technology integration process and revolutionary and evolutionary project outcomes. The following tables examine how the potential and yield estimates summarized in the previous section correlate with the processes observed during field work. Table 7-1 is drawn from the results of several regressions relating process and technological potential (see Appendix II at the end of this book). It indicates that effective revolutionary projects are associated with high experimental capacity and research experience, but not with project experience.

The analysis is consistent with the data on mainframe projects presented in Chapter 5 and indicates that effectiveness in revolutionary development is linked to knowledge-generation processes, such as experimentation and research. An association with knowledge retention through project experience is absent. This makes sense: If the change is revolutionary, much of the existing knowledge will be obsolete and the interaction between technology and

TABLE 7-1

Variables Associated with Effective Revolutionary Projects

Variable	Description
Experimentation capacity	Technological potential is positively associated with the capacity of the experimental setup. Differences in experimental capacity between projects is quite large, more than a factor of ten in several cases. Effective revolutionary projects were characterized by massive experimentation, exceeding 100 million trials per project.
Experimental iteration time	Not significant
Project experience	Not significant.
Research experience (internal and external)	Research experience was positively associated with technological potential. This was true for both external experience (that is, hiring Ph.D.s from universities) and internal experience (that is, experience in internal research laboratories).

environment will be unknown. To be effective, a project would need to generate knowledge about the interaction of technology and context as rapidly as possible. And much of this will have to be done through experimentation, since part of the experience with the application environment (that is, project experience) will be outdated.

Table 7-2, which describes the drivers of evolutionary development, is based on the results of regressions between process and technological yield. The results suggest that yield is related to factors substantially different from those correlated with potential. Project experience and the experimental iteration time are of primary importance.

The results show that experimentation is also important in evolutionary projects, although the experimentation strategy is very different from that leading to performance in a revolutionary situation. Rather than massive, parallel bursts of experimentation, effective experimentation in an evolutionary project is characterized by a targeted series of rapid experimental iterations. This effectively extends the knowledge retained through experience in a linear, incremental fashion.

Chapter 4 showed that organizational processes aimed at accumulating experience and performing experimentation were

TABLE 7-2

Variables Associated with Effective Evolutionary Projects

Variable	Description
Experimentation capacity	Not significant.
Experimental iteration time	Negative association with yield. The faster the iteration, the higher the yield achieved.
Project experience	Project experience was positively associated with technological yield.
Research experience (internal and external)	Not significant in any of the regressions.

associated with the timely introduction of new process generations. The analysis presented here adds a level of subtlety by indicating that the optimal balance of experimentation and experience depends on the technical approach of the project. The next two research cases reveal these differences in some detail. The first is drawn from the study of mainframe module subsystems and illustrates how NEC first built an impressive technological foundation through revolution and then leveraged it through evolutionary projects. The second, from the workstation study, shows how Silicon Graphics managed a series of revolutionary projects.

Technological Evolution at NEC

NEC's history of mainframe and supercomputer development is one of the most interesting examples of managing technological change effectively. In June 1985, NEC entered the supercomputer business by introducing the SX-2, whose impressive design and performance surprised existing supercomputer and mainframe manufacturers. The SX-2 was a revolutionary project, introducing a number of new technologies. While not the leader in its field, the SX-2 established an outstanding technological foundation for the future. The real payoff came with the next generation. In June 1990, NEC followed on the SX-2 project by delivering the first units of the SX-3, which became the world's fastest computer. The peak performance of the SX-3 beat that of the next generation Cray supercomputer (introduced over a year later) by more than a factor of two. NEC thus went from no industry participation to technical leadership in one of the world's most sophisticated industries in less than ten years.

The SX-3 was an outstanding example of *evolutionary* development. It built on the technological foundations created through the SX-2 project and increased the product's performance by a substantial margin without introducing any new fundamental technologies. The SX-3 was in all ways an extension of the SX-2. Its superior performance was not due to any fundamental breakthroughs, but rather to an incredible attention to detail in optimizing the match between technology and application context. (See Figure 7-2 for a time line.)

Revolutionary Foundations

The performance of the SX supercomputer series and of the related mainframe systems hinged on the design of their processor modules. These, in turn, centered on the development of polyimide materials, which greatly improved the electrical characteristics of traditional, ceramic-based module systems.

Advances in polyimide material technology can be traced to research efforts in the late 1970s. In 1977, conscious of the pressure on conventional ceramic materials that future performance

FIGURE 7-2

NEC Projects

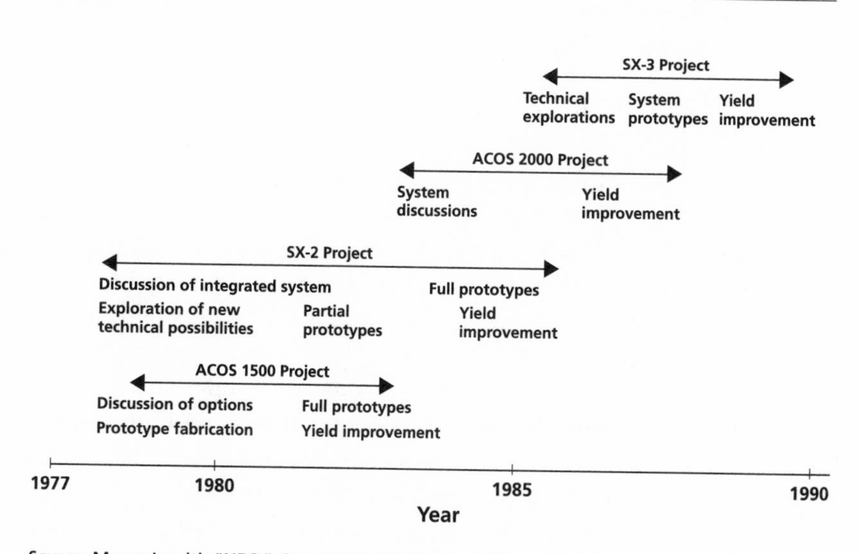

SOURCE: Marco Iansiti, "NEC," Case 9-693-095 (Boston: Harvard Business School, 1993).

requirements would bring, NEC scientists, in joint ventures with two material suppliers, began exploring the technical possibilities of polyimide materials for high density. The new materials had very attractive intrinsic properties that allowed for faster transmission of signals and higher density interconnection patterns.

Two groups within NEC were particularly critical to the SX series: the central research laboratory in Kawasaki and the integration group at the computer plant in Fuchu City. Members of the research group had been active in materials research for many years. The group comprised about half a dozen scientists with expertise in material science and several years' research experience at NEC. Most of their work was aimed at developing fundamental new techniques and approaches with application in novel systems. For its part, the integration group focused on creating a working, manufacturable product. While the role of the research laboratory was to offer a large breadth of technical options, the integration group would drive the investigations, select the most promising alternatives, and integrate them into a manufacturable subsystem.

In 1980, the integration group of NEC's computer division was hard at work developing the processor module for a mainframe computer, the ACOS 1500. The group comprised about a dozen young engineers. Some had been out of university only a few years, although many had been involved in the development of the previous generation mainframe, the ACOS 1000 system. Some members had also been actively involved in the polyimide joint ventures on a part-time basis and had strongly influenced the direction of the early research efforts.

SX-2 concept discussions started in 1980. The targeted shipping date was 1985, leaving less than five years for the entire effort. The first customer, a Japanese university, had already been identified and was involved in setting the computer specifications. During 1981, managers and engineers of the research and integration groups met repeatedly to discuss options for the supercomputer package. Members of the integration group naturally took leadership in setting the direction of the project and began to perform feasibility studies of new packaging concepts. They initiated discussions with members of the research laboratory and with material suppliers, system engineers, and integrated circuit designers. Many possibilities were discussed and modeled; the most promising were tested in rough prototypes. Integration group engineers and research specialists thus decided that four different concepts should be pursued in parallel. The first three were quite novel for NEC (two of these including polyimide materials), while the fourth approach was an extension of an existing concept.

Members of the integration group quickly began to perform feasibility studies for the different approaches. They initiated discussions with members of the central research laboratories, material suppliers, system engineers, and integrated circuit designers. They also investigated possibilities through experiments, quickly making many physical models of the module assembly, to investigate the characteristics of the new materials. One member felt strongly about the importance of subsystem-level models and prototypes:

> In my opinion, material performance can only be evaluated at the subsystem level. You cannot examine performance using elements. You must make subsystems using the material and then check things like reliability and functionality. It is the most effective way to test anything . . . basically make some substrate samples, build a fairly complete module, and see how it works.[3]

In October 1982, NEC senior management officially approved the SX-2 project and the level of resource allocation increased. The number of people involved from the integration group rose from five half-time engineers in 1981 to about a dozen engineers and ten technicians in 1982, with their level of involvement in the SX-2 project increasing as the ACOS 1500 project began to wind down. The research group and the polyimide supplier also dedicated significant resources to the effort, adding up to about 20 percent of the total. Between 1983 and early 1985, the allocation level of integration engineers and technicians remained about the same, although the team was gradually joined by about a dozen production engineers from the Fuchu City plant. As the effort continued, confidence in the polyimide materials grew.

During 1983, many models of the new packaging system were constructed to assess production yield; the emphasis was on completing partial prototypes and a number of models for each of several fundamentally different concepts. The yield of the polyimide thin films, for example, was investigated at each stage of the fabrication process, between the deposition of each layer. In this way, before the full prototyping process began, there was already sufficient data (on yields, types of defects, causes of defects, and so forth) to provide confidence in the eventual high yield of the process. Reliability testing began in November 1983 and was continued in 1984, leading to some changes in the composition of the polyimide. The first complete, fully representative prototypes were constructed in March 1984.

More than one hundred modules were built. A batch would be fabricated and tested until a major defect was encountered. Knowledge gained from analyzing the causes of the defect would then be used in the construction of a second batch of prototypes, while the first batch, now only partially functional, would be used to conduct additional tests. The integration group went through many such parallel iterations, an approach described as "thinking while running." The prototypes tested a wide variety of design options, and the design was gradually refined.

By mid summer 1984, the new module concept was complete, and the other three approaches were finally dropped. It employed polyimide materials, whose incompatibilities with the rest of the module system had been resolved. This involved the development of a number of specific technical refinements, including improved material compositions and many subtle improvements in the production process. By June 1985, the SX-2 concept was complete: The production process had been designed in detail, production workers had been trained, and the product had been shipped to the first customers.

From Revolution to Evolution

In early 1985, before the SX-2 project was completed, engineers in the integration group began working on adapting the new technical approach to the next generation of mainframes, the ACOS 2000, to be shipped at the end of 1986. Soon thereafter, the SX-3 project was also begun on an informal basis, with discussions about how the SX-2 concept could be extended to provide higher performance. Project engineers were often allocated on two projects (for example, the end of the SX-2 and the start of the SX-3) at the same time, facilitating the transfer of knowledge from project to project.

The ACOS 2000 was introduced on time at the end of 1986, exhibiting the fastest computer hardware in its industry segment. Its design was based completely on the SX-2, although the number of integrated circuits was reduced, given its lower performance targets (the ACOS 2000 was a mainframe, not a supercomputer). The focus of the engineering team was primarily on cost reduction. The ACOS 2000 was fully driven by engineers from the integration group responsible for the SX-2. This was still going at full speed at the time, with yield refinement and process improvement activities.

Integration engineers working on the ACOS 2000 project retained some responsibilities for the SX-2.

Early 1987 marked the official start of the SX-3 project. It was characterized by an approach similar to the ACOS 2000 and was staffed by many engineers with SX-2 project experience. By this time, several integration group members had participated in two project generations (see Figure 7-2), with responsibilities ranging from concept investigation to product introduction. The group's small size and the project managers' attention together provided members the chance to obtain broad exposure to a wide range of tasks. Their approach involved project members' continuous participation in the development project; there were no hand-offs in the middle of the effort. As a result, integration group members had built an intimate knowledge of the product's context, represented by existing production processes and system-level design considerations. This put them in an ideal position to analyze the interaction between the new approaches and the organization's existing capability base and infrastructure.

Such level of experience in the integration group was reflected in many of its decisions. These ranged from relatively major changes such as a redesign of the cooling system to account for the higher performance specifications to more subtle choices such as the decision to define the pitch between chips to be the same as the pitch between pins at the bottom of the substrate, to allow for vertical connections and minimizing wiring length and signal power.

For both the SX-3 and the ACOS 2000 projects, experimentation was much more limited in comparison to the SX-2. The idea was to validate the designs based on the experience of the integration group members, not to explore fundamentally new possibilities. Experimentation efforts were highly representative, however, since they were carried out in the same facilities and by the same technicians responsible for building commercial SX-2s.

The results were impressive: While the SX-3 retained essentially the same basic technologies developed in the SX-2, the new packaging module included forty times the number of transistors as the earlier model. The SX-3 was introduced in 1990, surpassing existing supercomputer models in performance and cost. The experience of the integration engineers, their intimate knowledge of the manufacturing system, and the high skill of the technicians all contributed to a design that met aggressive performance specifications and was delivered on time. The SX-2 project set up a strong technological platform for NEC. In the ACOS 2000 and the SX-3, this platform was extended through evolutionary efforts that enhanced the performance of the systems and established NEC as a leader in the supercomputer industry.

Revolution Follows Revolution at Silicon Graphics

When SGI introduced the Challenge server in the spring of 1993, the product was an incredible success; its revolutionary bus design enabled it to set the record for the fastest transaction processor in the world, and its sales came close to $1 billion in the first year. Ironically, one of the management team's first decisions after the product's introduction was to go back to square one and completely scrap its technological foundations. Revolution followed revolution at Silicon Graphics. The resulting product not only reestablished SGI's leadership at the high end of the server market, but the scalability and modularity of its novel architecture also enabled a completely different product line strategy at SGI, extending the company's reach into entirely new segments of the market.

In the six months following the Challenge's introduction, a small group of engineers started working on a completely different type of architecture. These efforts started a project called Lego, which led to the world's first distributed shared memory commercial server, introduced in the fall of 1996 (see time line in Figure 7-3). The project provides a wonderful illustration of revolutionary development in the server and workstation environment.

University Foundations

Lego had its foundations in university research. The distributed shared memory (DSM) concept had been discussed theoretically in academic circles for many years. And by the late 1980s, several universities began significant projects to build functional physical systems. The essence of DSM is relatively simple. The architecture of a traditional computer centers on a "bus," the central conduit of information linking all components and subsystems: the processors, the memory, the storage media (disk drives, CD-ROM, and so on), and the input/output devices (screen, keyboard, and so on). This type of architecture (shown in Figure 7-4 (a)) has characterized essentially all commercial computers sold in the last thirty years, from high-end mainframes to personal computers. Its advantage is that it is simple and flexible. Its main disadvantage is that as the speed of the component increases (particularly processor

FIGURE 7-3

Challenge and Lego Projects at Silicon Graphics

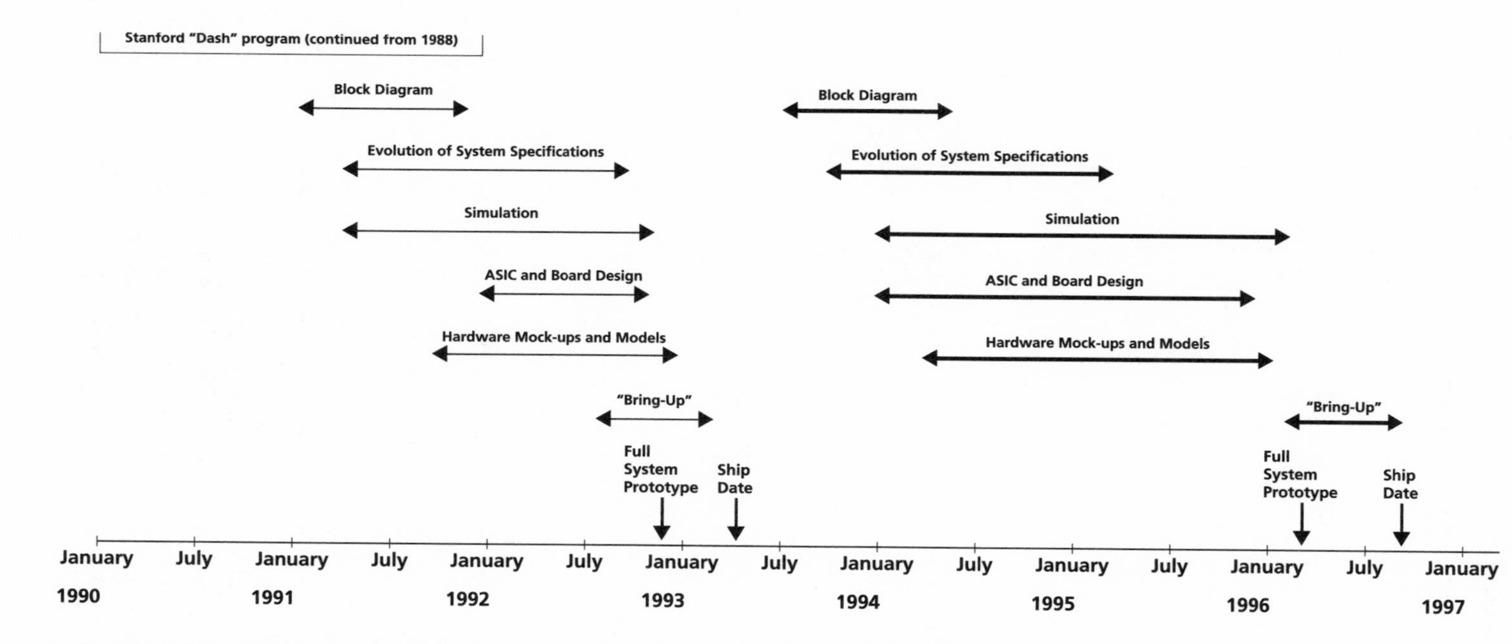

NOTE: Events marked by thin arrows indicate Challenge project; events marked by thick arrows indicate Lego project.

SOURCE: Challenge project information adapted from Ellen Stein and Marco Iansiti, "Silicon Graphics, Inc.," Case 9-695-061 (Boston: Harvard Business School, 1995).

FIGURE 7-4

a. Traditional Bus Architecture

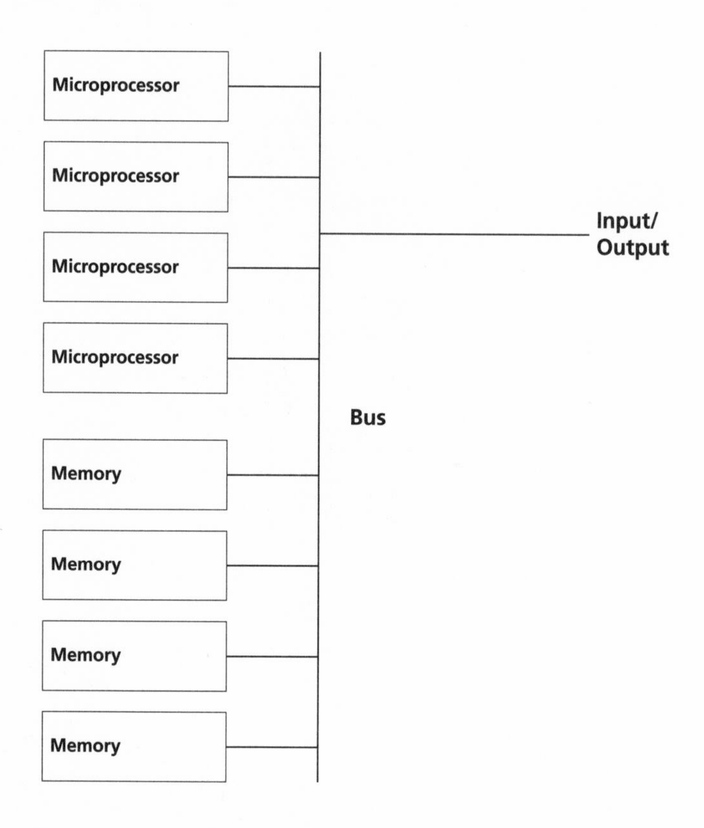

and memory), the bus tends to become a bottleneck in machine performance.

The idea behind the DSM architecture (shown in Fig. 7-4(b)) is to resolve the bottleneck problem by eliminating the single bus structure. This new architecture groups bundles of logic and memory together in integrated modules. These are connected to each other in a star configuration that comes together at a router, which controls the information traffic in the computer. This architecture has several advantages. First, computer speed is increased significantly, since critical instruction sequences are confined to the integrated processor-memory modules and information need not continuously travel on the bus between independent processor and memory. Second, the design is highly scalable: Since there is no bus to act as a bottleneck, memory processor modules can be added with

b. Distributed Shared Memory Architecture

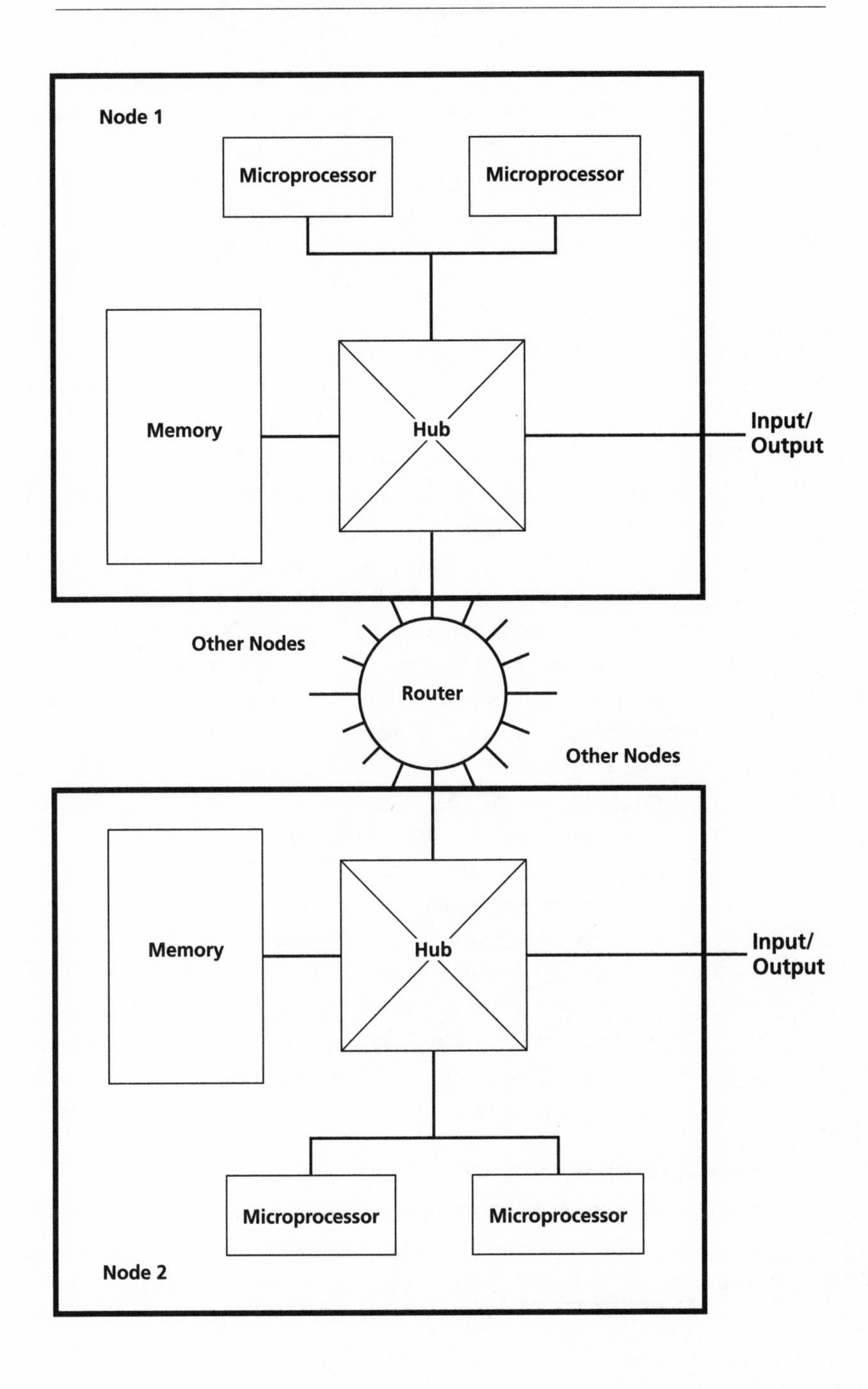

almost no limit. Third, the architecture is very reliable, since errors can be confined to individual modules.

The universities most active in exploring the distributed shared memory concept in the 1980s were MIT, the University of Wisconsin, and Stanford. The group at Stanford was led by John Hennessy, a professor of electrical engineering who had been involved with Silicon Graphics on many previous occasions. He assigned several graduate students to a DSM research project called Dash. This started in 1988, with the objective of creating a functional prototype by 1991. Jim Lauden, David Nakahira, and Dan Lenoski were among this group.

Before getting his Ph.D. at Stanford, Lenoski had worked for many years at Tandem Corporation. With extensive experience in ASIC and hardware design, he had participated in several product development efforts. As he recounts the story, "I was useful in the Dash program because I had demonstrated I could build things."[4] By the time he came to Stanford, Lenoski was in an ideal position to contribute to DSM research.

The Dash program was performed as basic academic research in computer science, conducted with outside funding. Still, Hennessy's connections to SGI significantly helped get Dash going. The Stanford group started with a Silicon Graphics Power 4D/340 workstation (a four-processor system, the precursor to the Challenge server) and frequently used SGI facilities to design circuit boards and debug the evolving system. Silicon Graphics engineers also provided advice and feedback to the Stanford team. As Lenoski recalled, "They were very open, we used their lab space, and some of their technical experts really helped out. They even built boards for us."

The design started by building on the 4D/340 system directly. The team took out one of the four processors and substituted for it a chip interface that would enable it to connect to other systems without going through the bus. Doing so allowed them to attach another module to the original system. The scheme was repeated in a daisy-chain until, by 1991, the computer had grown to include sixty-four processors.

The Lego Project

Interest in the DSM architecture at Silicon Graphics grew as the Stanford team continued working on its prototype. At the time, however, the Challenge project was at its peak, and engineering resources in the high-end group were dedicated to its execution. Actual work on DSM processing at SGI began in 1993, when the company hired Jim Lauden, one of the recently graduated Ph.D.s from the Stanford team. Lenoski followed Lauden soon thereafter.

SGI was interested in the DSM architecture for several reasons. First was the relentless drive for increased performance. In the Challenge project, aggressive targets had been achieved by working intensively on the bus, increasing its capacity to transmit information by more than a factor of twenty. This had been very difficult to do, and the possibility of bypassing the bus altogether was very attractive.

Other critical reasons included cost and design flexibility. The scalability of the DSM architecture removed one of the highest cost items from the design—the bus itself. If relatively inexpensive modules, each containing only a small number of processors, could be designed, it might be possible to create an architecture for a system that would serve a wide variety of workstation segments, from midrange workstation customers to commercial server applications for database processing to high-end supercomputing.

Finally, the DSM approach was selected because of its potential for higher reliability. The Stanford experience had established the feasibility of erecting fire walls between the different system modules. This would help predict and contain errors in the system. High reliability would be extremely attractive to commercial customers.

While the reach for increased performance was natural for SGI projects, the focus on modularity, cost, and reliability improvements would enable a significant strategic shift. Traditionally, SGI had emphasized sales to engineering customers.[5] Lego might enable SGI to expand considerably its presence in commercial segments, such as telecommunications and banking, since its architecture and design approach appeared flexible and reliable enough to really fit their needs. Several senior managers got involved in the early stages of the project to influence its direction. SGI's chief technical officer was a key champion of the DSM approach. Other senior managers were more skeptical of the viability of the new architecture, but encouraged the investigations, as long as the clear focus on low cost and reliability was kept. The experience at Stanford became essential to SGI's eventual acceptance of DSM and the success of the project.

Dan Lenoski and Jim Lauden brought with them a clear idea of what SGI would need to do to implement the new architecture, and their experience with distributed shared memory systems was critical in convincing SGI management of the feasibility of the new approach. Other external resources were combined with the Stanford expertise. Given the high frequency of the connection between modules (values around 400 MHz were being suggested), SGI set up a signal integrity group to investigate the feasibility of long-range signal transmission. The resources for this group were drawn largely from people hired from IBM, which was then downsizing its high-end operations. Their expertise was combined with that of SGI "insiders" to define the Lego concept. Rick Bahr led the hardware part of the team while Rich Altmaier led the software effort. Both had been protagonists in the Challenge project. Mike Galles, a principal engineer extremely experienced in ASIC design, developed the critical router chip, which dramatically extended the potential of the Stanford design. At the senior level, managers continued to have influence on the project as it began to evolve. Wei Yen, executive vice president in charge of high-end systems, kept pressure on cost, encouraging the team to develop increasingly small and inexpensive processor-memory modules.

By January 1994, the project included about half a dozen full-time people. By late spring, the team had increased to almost its full complement, and confidence in the new architecture was increasing. By year end, the basic configuration of the router and hub design was down.

The Lego project was driven by massive experimentation, which took several forms. The team built some critical hardware models to test out the system's feasibility at high frequency. In late 1994, a critical test vehicle finally confirmed that 400 MHz operation was feasible. The simulation effort also began to gain momentum during 1994, reaching its peak by March 1995. The total simulation power at SGI's disposition was enormous: They were running a total of two hundred twenty high-performance processors. Load-balancing software ensured that these were used efficiently. Alex Silbey, who ran the simulation setup, estimated that the hardware used in developing Lego ranked in the top one thousand supercomputing centers around the world. This power was used to run several types of simulation models, from high-level behavioral models down to the actual routings in the semiconductor chips.[6]

This massive experimentation led to several architectural changes in the project. Major redefinitions occurred in August, September, and, finally, in early November 1995. By late November SGI finally locked into the design, and the first system boards were delivered in March 1996 to start the bring-up stage. The results were impressive. A week after the system was turned on it was capable of running a version of UNIX, despite the aggressiveness of its design. Lego shipped to great reviews in the fall of 1996.

Contrasting Evolutionary and Revolutionary Management

The data from the semiconductor study can be combined with the two examples above to illustrate the drivers of evolutionary and revolutionary development. Table 7-3 summarizes the key differences by referring to the case examples described.

Revolution

Effectiveness in revolutionary projects is associated with research experience and experimentation. The introduction of novel technology is linked to an approach that cultivates fundamental, specialized knowledge and transfers it into the project through individuals. This knowledge is merged with the context of application through a

TABLE 7-3

Contrasting the Characteristics of Effective Evolutionary and Revolutionary Projects

Variable	Revolution (Lego)	Evolution (SX-3)
Experimentation	Massive simulation capability; mockups and models used as well. Diversity of tools and accuracy of representation at the system level critical.	Selective prototyping carried out to validate experience. Rapid iterations. Prototyping performed in the plant by production technicians to ensure representativeness.
Project experience	Less than 50 percent of the project members involved in the Challenge, the previous server generation.	High; all key project members had participated in the development of the SX-2.
Research experience (internal and external)	Critical transfer of knowledge from research performed at Stanford University. SGI hired about half of the Stanford research team. Knowledge also transferred from internal research and efforts at other companies (for example, IBM).	Not significant in SX-3. Research had been critical in SX-2.

large number of experimental trials, which generate new information about technology-context interactions.

The amount of experimentation is truly massive. Projects emphasizing this approach can run millions of trials during a project, investigating a broad set of technical alternatives. These are compared against each other, with a relatively well-defined output target in mind. In the semiconductor study, this experimentation came in an intense, parallel, but relatively short burst. As such, there was ample time to transfer the novel technologies to the manufacturing setting. The workstation environment, in contrast, did not require such long lead times for the bring-up stage. SGI engineers claimed that their simulation capability was essential to ensure the effectiveness and speed of the later project stages. In a relative sense, however, comparing Challenge to Lego, for example, the more revolutionary project emerges as requiring a longer bring-up stage.

In all environments, revolutionary projects were characterized by a strong input from research. The research could come from a variety of sources, including internal research laboratories, external suppliers, competitors, and universities. There did not seem to be a direct correlation between project performance and the funding or support of research, however. In the semiconductor study, direct research expenditure was actually inversely correlated with project performance. Rather, real impact came from the transfer of individuals, usually through permanent hiring. This implies that the knowledge required to implement the novel, revolutionary technologies is difficult to articulate and is best carried by direct experience. In effective organizations, new project members were frequently chosen to bring in critical research experience from universities. The leader of one of the most effective semiconductor projects explained:

> There are several university departments that we monitor very closely. We know what students are working on, and when they are graduating. We will choose a few of these students that have worked on technologies that are important to us and target them for hire. Hiring these hand-picked graduates is a critical source of new knowledge for us.[7]

The results here, therefore, do not imply that research is not a valuable activity. They do suggest, however, that the value of research can be wasted if the knowledge generated is not transferred through individuals.

While this study found no projects without a significant experience base, depth in *project* experience was not the primary factor in differentiating between more or less effective revolutionary projects. This can be understood in part by the obsolescence of context-specific knowledge in revolutionary situations. Revolutionary projects overturn the established context. It is thus unsurprising that detailed knowledge of this context is not critical to bringing new technologies to the market. Performance in revolutionary projects was linked to knowledge creation rather than to knowledge accumulation.

Evolution

As NEC's SX-3 example makes clear, large performance increases were possible through evolutionary as well as revolutionary approaches. Effectiveness in evolutionary projects is associated with a deeply contrasting approach, however. The most important factors are project experience, the similarity between experimentation facility and the application context, and the timing of experimentation activities. Rather than using massively parallel experimentation, effective evolutionary projects relied on a sequence of carefully selected experiments, gradually optimizing the match between technology and the system context. Moreover, the time window for experimentation appears longer in this case. This leaves more time for optimizing the technology-system interactions. Transferring new technologies to the manufacturing setting does not require as much time, since the projects are making extensive use of existing, mature equipment.

Project experience is essential for effective evolution. Experienced project members drove the SX-3 project, obtaining a dramatic increase in performance by stretching the existing approach through innumerable refinements. When asked where the ideas for the work came, one project manager clearly noted the linkage to the previous project:

> The ideas for these improvements came from the SX-2 project, almost without exception. I clearly remember being frustrated during the SX-2 that it was too late to make the changes that we wanted to make and thinking that at least we would have a chance to do so on the next effort.[8]

The rich knowledge of experienced project members was also the key driver in selecting the limited number of experimental trials.

The semiconductor study involved discussing this issue explicitly with the managers of more evolutionary efforts. When asked why their project had so much less experimentation capacity than did competing projects, one of these managers answered:

> Why would you want to have so many experiments? We wouldn't know what to do with all that data. No, the important thing is to run the right experiments, not to try to do everything blindly. We spend a lot of time thinking about which experiments should be run. This is where our experience base is most useful.[9]

In the most effective evolutionary projects, the optimization of technology choice goes on as late as possible, indicated by the organization's commitment to the architecture of the production process.

The evidence implies that managing a project's experience base is important to performance. Evidence also suggests that some level of project turnover may be critical, even in evolutionary situations. Several project members elaborated on this point, stressing the importance of bringing in new experience and perspectives. One member described the group's goals in this area:

> We try to bring in about 10 percent new project members every year. This achieves two things. First, we get enough new people trained so that there is a solid base for future projects. Second, it brings in new ideas, and makes sure we don't get too inbred.[10]

In summary, effective evolution is linked to mechanisms different from those driving revolution. Rather than an exhaustive focus on knowledge creation through research and experimentation, evolution appears driven by a careful balance. The focus is on knowledge accumulation through experience. Gaining some new ideas through new project members and having a good setup for representative experimentation, however, are important to extend old concepts in effective ways.

Evolution, Revolution, and Inertia

Many authors have argued that a central obstacle to the effective management of technology is knowing when and how to break with the past. This judgment involves balancing the efficiency of routinization against the dangers of the associated organizational inertia. The empirical environments chosen for this research accentuate the inertial problem, since complexity creates the need for routinization while novelty amplifies its danger. The evolutionary and revolutionary approaches discussed in this chapter can be understood as different strategies for meeting these challenges.

Revolutionary projects aggressively attack inertia. They invest in research and investigate a broad range of options through massive, parallel experimentation. An effective revolutionary project thus focuses on knowledge generation to renew outdated routines. This is done for two reasons. First, the project cannot bank on knowledge retention since much of the existing knowledge is obsolete. Second, research and massive experimentation proactively break inertia, essentially forcing the organization to face up to the impact of new technological possibilities. This sparks what various authors have called "second-loop" or "second-order" learning processes (see Argyris and Schon, 1978; Levitt and March, 1988; and Lant and Mezias, 1992). These processes go beyond the refinement of existing routines to the creation of new frames of reference, such as the distributed shared memory approach in Lego.

In deep contrast to the "brute force" approach followed by projects emphasizing revolution, those following an evolutionary path manage inertia through a careful balancing act. On the one hand, they maintain a critical mass of experienced engineers who really know the product and production system. On the other hand, they periodically include new talent to break with past trajectories, bringing in new ideas. Experimentation also plays an important but different role from the one it plays in revolutionary projects. In evolutionary settings, experimentation is selective, guided by project member experience, and used to validate decisions—as was done for the SX-3 project. These projects therefore appear to emphasize a first-loop learning processes (see Argyris and Schon, 1978), relying on the optimization of existing equipment, procedures, and routines. First-loop processes do not suffice, however. Even while following an evolutionary approach to managing technological change, it is important to perform experimental iterations to evaluate new approaches. Fighting inertia, therefore, is also critical in a project with a strong evolutionary component. Rather than through the aggressive approach associated with a revolutionary path, however, inertia is fought by carefully meshing experience and experimentation in an effort to leverage mature technologies without overextending them.

The evidence gathered for this study has important implications for the management of technological change. It implies that the technology integration process is a critical window of opportunity for recognizing and responding to technological threats and opportunities. Additionally, it suggests that the nature of organizational response is at least partly under the control of managers, not only through specific technical decisions, but also through the organizational processes they implement. Managers can thus build capability for managing technological change by defining the structure of the technology integration process, the experience base employed, and the experimentation tools available.

The next two chapters present further evidence to substantiate this argument by discussing how several organizations have built technology integration capability. These chapters show how the restructuring of R&D organizations with a focus on technology integration, the management of individual experience, and investment in experimentation can be linked to striking performance improvements.

Notes

1. This application of technological potential to assess the revolutionary nature of a project has some important subtleties, however. Most critically, the work in this book generally assumes that the performance model used in the calculation of technological potential is invariant in time. This is often, but not always, the case. Most critically, when market requirements change, a similar change in the performance function is to be expected. This would be the case in the disruptive technology examples described by Christensen (1997). The methodology here still applies, however, and can account for such changes by making the performance function time dependent.
2. This approach allows the possibility that a project has both high evolutionary *and* revolutionary components. This is not unreasonable, since the systems that are being developed are complex, and while one part may evolve in a discontinuous fashion, another part may evolve in a more gradual fashion. The data suggests that the situation occurs only rarely in practice. Very few projects here show high components of both potential and yield. This might have technological roots; once a part of the system evolves in a discontinuous fashion, it will influence many other parts, since the system is complex. Additionally, this effect may also have organizational roots, implying that it is difficult to manage revolutionary and evolutionary change simultaneously.
3. Marco Iansiti, "NEC," Case 9-693-095 (Boston: Harvard Business School, 1993).
4. Interview conducted by author during field research from 1990 through 1995.
5. SGI had begun to target some commercial customers with the Challenge project, which preceded Lego, but its penetration was limited by the product's characteristics and design approach.

6. By dedicating fourteen processors to the task, the system could be simulated at a speed of one cycle per second.
7. Interview conducted by author during field research from 1990 through 1995.
8. Ibid.
9. Ibid.
10. Ibid.

Building Technology

chapter 8

Integration Capability

IN 1990, SEVERAL ACADEMICS and practitioners predicted the imminent demise of the U.S. computer industry (see Rappaport and Halevi, 1991). American firms were in retreat on all fronts: IBM had lost substantial market share to Japanese manufacturers in every hardware segment; Intel was consistently late in introducing new generations of semiconductor technology and new chip designs; even Microsoft's competitive position in software was jeopardized by severe product introduction delays and reliability problems.

A scant five years later the situation was very different. U.S. firms had regained lost ground across critical segments that included semiconductor components, personal computers, servers, and laptops. A handful of Silicon Valley–based firms dominated the workstation industry. Intel and Microsoft had consolidated their leadership in microprocessors and software. IBM had improved dramatically its development and manufacturing capabilities, introducing impressive new products across its varied businesses. A fresh generation of U.S. startups like Netscape and Yahoo! had staked out the latest growth segment, Internet software. This study was ideally suited to observe what drove these dramatic changes in product development performance and industry evolution.

This book so far has argued that technology integration is important to performance. It has shown in some detail the mechanisms underlying an effective technology integration process and their implications for managing revolutionary and evolutionary projects. In the next two chapters, I take the argument a step further by analyzing this striking turnaround in the U.S. computer industry and revealing how organizations can build technology integration capability over time. To do this, I will examine detailed case histories documenting significant changes in organization and process and showing associated improvements in performance.

The focus of this chapter is on the semiconductor industry, a classic science-based environment. The large investments in R&D and capital equipment, the reliance on advances in solid state physics and materials science, the significant role played by traditional industrial labs like IBM's T.J. Watson Research Center and AT&T Bell Laboratories—all these practices and institutions associate the industry with an almost stereotypical emphasis on technology as a competitive weapon. But these traditional institutions and practices had to change, and do so deeply, to respond to the competitive threats of the 1980s and 1990s; and, as this chapter will show, building technology integration capability was central to this effort. Chapter 9 extends the discussion to personal computer and Internet software, environments completely different from semiconductors. Companies in these environments do not rely on deep traditions in research laboratories. They do not invest large amounts in research or in manufacturing infrastructure. The complexity of their challenges arises from a rapidly evolving customer base and network architecture rather than from a massive manufacturing infrastructure. Nevertheless, as Chapter 9 reveals, product development capability is related to some of the same central ideas.

The Turnaround of the U.S. Semiconductor Industry

In the 1960s and 1970s, U.S. firms first invented and then dominated the semiconductor industry. AT&T Bell Laboratories was responsible for a wide variety of innovations, from the transistor to electron-beam lithography. IBM was the first firm to develop a truly high volume semiconductor production process, for its transistor-based "solid logic technology," introduced in 1964. High-volume semiconductor production was later perfected by Texas Instruments, while most popular integrated circuit logic and memory designs were first introduced by U.S. firms like Fairchild and Intel. During the 1960s and 1970s, U.S. firms were responsible for the lion's share of the market in virtually every segment of the semiconductor market, from discrete components to integrated circuits.

The 1980s, however, witnessed the rise of powerful new semiconductor manufacturers in Japan and Korea. Over a brief period, a group of Japanese firms led by Hitachi, NEC, and Toshiba gained a substantial advantage in the development of new process technology, investing heavily in technology integration and manufacturing capability. These firms were soon joined by a group of Korean firms, headed by Samsung, which achieved market share leadership in DRAMs by the early 1990s. U.S.-trained scientists and engineers brought knowledge back to Asia, transferring the latest concepts in lithography, etching, and transistor design. The underlying fundamental science and technology base thus diffused from its birthplace in the United States to emerging countries like Korea and China.

U.S. and European firms fell years behind in the development of process technology, which put them at a tremendous disadvantage, particularly in the DRAM business, where most of the profits are made the year after a new generation of process technology is introduced. Prices fall rapidly after the first DRAMs hit the market, and a six-month project delay can mean the difference between enormous profits and significant losses. By the early 1980s, Japanese firms were already in the lead. IBM, Texas Instruments, and Intel introduced the 256K DRAM generation more than a year and a half behind the leader, Hitachi; by the next generation (1 megabit), the problems for U.S. firms worsened, and they were very far behind. At this point, Toshiba, NEC, and Hitachi were in control of the market, with Samsung gaining ground. Intel, National, AMD, Motorola, AT&T, Mostek, and INMOS soon exited the DRAM industry.

When asked about the root causes for these dramatic failures, academics and practitioners often cite a lack of investment in manufacturing capacity (using arguments based on the short-term focus typical of U.S. business or on Wall Street pressures for profitability). When I analyzed these projects in detail, however, I discovered that delays had occurred long *before* massive manufacturing investments were necessary. It was also clear that the density performance of U.S. products was inferior. Die sizes were consistently larger than those of their Japanese competitors, which led to huge cost disadvantages. The problem, I discovered, did not arise from the size of R&D dollar investments or capital expenditures, but rather lay with the process development and manufacturing capabilities of these firms. Facing the prospect of investing massive amounts of money in a process that would get to market late with an inferior product, U.S. managers, not surprisingly, declined to do so!

Yet by continuing to follow the industry, evidence emerges of an impressive turnaround by those U.S. manufacturers deciding to remain in the business. Figure 8-1 shows the DRAM segment and graphically depicts a striking improvement among non-Japanese competitors. The figure, which focuses on the most critical measure of integrated circuit performance, its density, displays the residuals of a regression of the logarithm of gate density against time, which takes out the long-term trend in the industry. A difference of *1* on the figure corresponds to an advantage of about a year and a half in the introduction of comparable products. (A value greater than one indicates a higher than average performance; a value less than one indicates inferior performance.) The data are aggregated by DRAM generation (from 64 kilobit to 64 megabit) and by country of origin.

Consistent with the argument made earlier, the figure shows a very substantial performance gap between Japan and the rest of

FIGURE 8-1

DRAM Performance

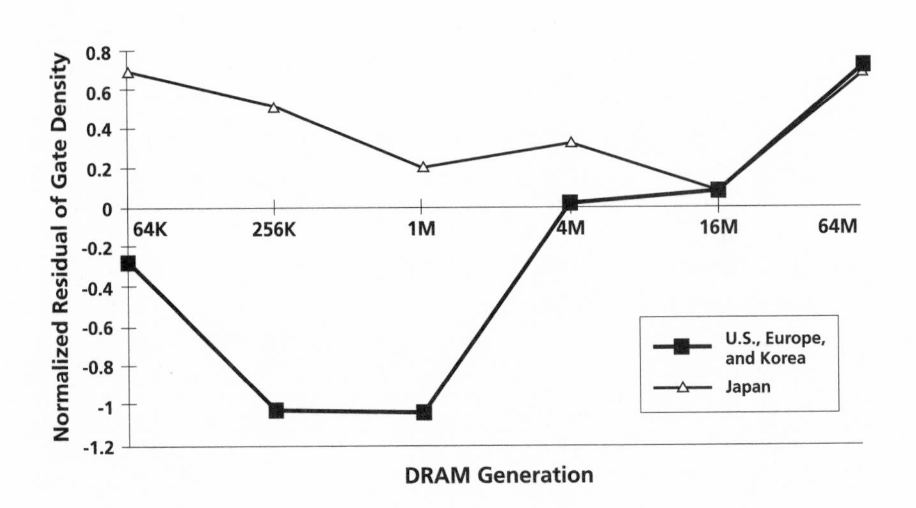

the world, between the 64 kilobit to 1 megabit generations. At its maximum, the gap adds to more than a two-year average delay in the introduction of a comparable product. The figure also shows, however, that the gap narrows in the 4 megabit generation and is completely erased by the 16 megabit and 64 megabit generations. The Japanese organizations indeed had enormous lead time and density advantages in the generations completed in the mid and late 1980s, but this advantage completely evaporated in the 1990s.[1] In part, this is because some of the worst competitors exited the industry (see Chapter 4). Several other competitors, however, significantly improved. (IBM's performance, for example, follows the non-Japanese trend line almost exactly.) Additionally, the figure shows that for the last generation both country averages are significantly above the trend. Since the trend is estimated from data on all generations, this indicates that the remaining competitors became faster. Firms still in competition had thus raised the average rate of technological progress.

Figure 8-2 shows a similar improvement trajectory for projects in the logic environment. The residual plot is aggregated differently in this case. Microprocessor logic products have a much higher design content than DRAMs, so firms can get away with inferior process technology for a while longer. Therefore, to show the

FIGURE 8-2

Microprocessor Performance

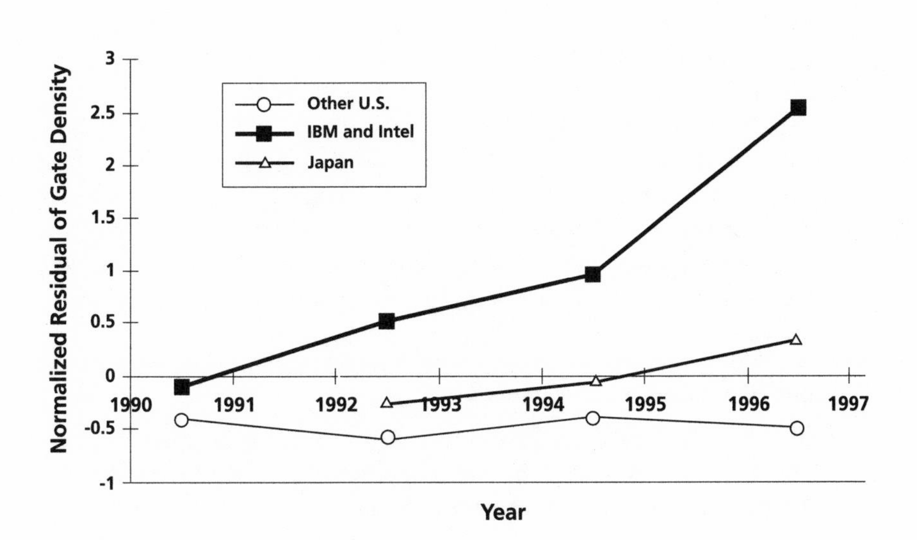

improvement in leading firms, I divided the U.S. competitors into two groups. The first, composed of IBM and Intel, dramatically increased project performance; the second, a more numerous group, continued to lag. The difference between the leading group and the others is quite large in 1996, corresponding to several years.

By the mid 1990s, these substantial performance differences had begun to undermine the second group of organizations. AMD attempted to address its profitability problems by merging with NexGen, which had access to IBM process expertise through an alliance. Similarly, Cyrix, Motorola, and Hewlett-Packard decided to enter into alliances and subcontract most of their process development and high-volume microprocessor manufacturing, the first two to IBM and the third to Intel. By 1996, only DEC remained as a truly independent competitor in the less-efficient U.S. group. The message from the observations in the logic environment is thus consistent with that found in DRAMs: Competitors that do not learn how to introduce new process generations early exit a competitive environment.

Improvements in U.S. Firms

Thus several semiconductor manufacturers achieved a striking improvement in performance during the early 1990s. Building technology integration capability was central to this turnaround, as indicated by the field data set from the study, which includes details on most major competitors in the industry. The story of Intel shows how this happened, as the result of numerous changes implemented by the company. Additional research by Robert Burgelman provides further details of the evolution of Intel's process development capabilities during the 1980s, and of the problems that led to their departure from the DRAM industry.[2] Burgelman also describes how Intel's top management used the lessons they learned from their experience in the DRAM business to engage in "creative destruction" of Intel's traditional approach to integrating process technology development and manufacturing.[3] Following Intel's story, this chapter shows that similar responses were made by a broader group of U.S. firms, including IBM and Texas Instruments.

Intel's Improvement Trajectory

In the early 1980s, Intel's technology integration in process development was conducted by many groups working independently.[4] These were scattered in business focus across DRAMs, programmable memories, and microprocessors; they were scattered in function among research, technology development, and manufacturing. Each production fab manager had final authority in purchasing each piece of equipment, and experimentation capacity was among the industry's lowest.

Today, process improvement is conducted centrally. A single development fab is transformed from an enormous laboratory into a volume manufacturing plant for each major generation. Experimentation capacity is massive, now the industry's highest. Technology and equipment choices are controlled by a single organization, part of the technology development group and under the direct scrutiny of Intel's executive vice president in charge of both process development and manufacturing. Once a process is established, it is replicated at fab after fab.

These changes were not accomplished in a single stroke. The story of process development at Intel provides an intriguing example of the development of technology integration capability. It begins around 1983, when Intel's production was split between three major product lines: DRAMs, a high-volume, low-margin product; EPROMs and EEPROMs, specialty memory products; and microprocessors, a high-margin product that represented an increasing share of the firm's revenues. Three geographically and organizationally distinct development efforts supported these lines. Although most manufacturing equipment was similar across them, the critical technologies for each product line differed in important ways—because technology choices had been made independently by a variety of groups.

Although each group was focused by business responsibility, each in fact drove a subset of technology choices that were partially replicated across the company. Technologists in the Santa Clara, California, group, which was responsible for programmable memories (EPROMs), emphasized processing techniques unique to the "floating gate" characteristic of those devices. Technologists from the microprocessor group in Livermore, California, concentrated on metallization and layering. At the same time, engineers from the Portland, Oregon, group drove the direction of lithography technologies that were critical to DRAMs.[5]

Top management tried to promote firm-wide equipment standardization by holding periodic meetings at which development strategies were discussed; the three process groups agreed to common sources of equipment when possible, but their perspectives were usually impossible to integrate. It was intended that differences between production lines be exploited as well, however. Specialized needs in the manufacturing processes of one line of products today would often drive improvements in other lines tomorrow. Because the DRAM business was largely driven by cost competition, for example, DRAM developers led efforts on advanced lithography used to shrink circuits to occupy less silicon. When the necessary tools had been proven in DRAM production, they were incorporated in the other product lines, which could be as much as one product generation behind DRAMs in the adoption of advanced photolithography.

The consequence of this system was that technology choices were carried out in a scattered and opportunistic fashion. Microprocessor developers would face lithography decisions that had been greatly influenced by the DRAM group, which operated under different constraints. Meanwhile, DRAM manufacturers would be pressured to use metallization technology choices made for a different context. The result was a process that did not work together coherently and that contributed to substantial problems and associated delays. And, once the process was finally transferred, plant personnel would have to change it repeatedly to achieve acceptable yields, often ordering different equipment—once again leading to delays, investment overruns, performance problems, and further decreases in standardization.

The lack of experimentation facilities in development only worsened the situation. Before 1985, process development groups were small, consisting of only a few engineers and technicians, who shared fab space and production capacity with manufacturing operations. The situation was described by a member of the DRAM process development team at Intel during this period:

> Back then we used a corner of the DRAM fab to do limited fabrication of process modules. The role of the development group was to produce a yielding process, not an optimized one, before handing it over to manufacturing.[6]

Because of this approach, only limited numbers of samples were tested by technology development before being transferred to the manufacturing organization. For example, during the development of the "one micron" process technology that was completed in 1989, Intel engineers "never did volume manufacturing in development."[7] These limits in experimentation capacity in turn led to limits in the scope of experimentation and testing. Efforts were focused on individual manufacturing modules, such as sputtering or optical lithography, and little attention was paid to developing a process that worked well as a coherent system. As an Intel engineering manager involved in these efforts noted, the approach achieved "little integration."[8]

These problems were exacerbated because technology developers were dispersed across a variety of locations, fragmenting their combined expertise. Therefore, when technology development was ready to transfer a process to manufacturing, the process was far from ready to reliably deliver results at high volumes. Worse, the yields achieved immediately following the process transfer were typically inferior to those achieved during development, and led to a characteristic U-shaped curve of yield versus time when plotted out. To improve post-transfer yields, manufacturing engineers would spend considerable time optimizing the manufacturing process and might even redesign a significant portion of the equipment. As one senior manager in the technology development organization explained, "[In the] move to a production factory, [we would] choose new equipment and wafer size, redesign core modules to fit existing equipment, [as well as make several] process flow redesigns."[9] In the 1.0 micron generation, for instance, it took twenty-one months for engineers in the second production fab to achieve yields already achieved in the first fab.[10] The same manager explained that this was not unusual: Intel had "basically a two-year process from end of development to production. [Additionally,] utilizations never matured, [and] every factory was unique."[11]

In many instances, Intel attempted to use innovative solutions to work around basic deficiencies in its process for technology selection and development. These attempts often backfired, however, causing even greater delays. During the 64K DRAM generation, for instance, Intel was the first implementor of on-chip "redundancy," while most other firms were emphasizing traditional avenues of cost reduction and process improvement.[12] Intel's approach involved the fabrication of additional circuits on the DRAM chip that, during testing, would automatically replace any defective circuits identified. However, problems in implementation caused the redundancy strategy to backfire late in development,[13] contributing to a significant delay in shipment—a delay estimated to be almost a year.[14] This unfortunate delay came at a critical time when other firms were building enormous capacity. As a result, Intel never achieved a major share of this market.

Like many other U.S. firms, Intel tried to circumvent these process development problems by being even more aggressive in design. After all, it had invented all the major product lines it sold.

Intel would differentiate itself from competitors by offering superior features for a price premium, which would attempt to mask problems in process development and manufacturing. In order to provide improved power consumption and speed, for example, Intel made the technological transition from NMOS to CMOS DRAMs for the 256K generation. This was one generation ahead of the industry, at a time when NMOS was still the cheaper technology. Once again, however, Intel was very late to market, a situation exacerbated by the design challenges. Soon the price of conventional NMOS DRAMs dropped so low as to obscure CMOS DRAMs' advantages and obliterate Intel's margins. Because these problems were systemic, and even though it had invented the DRAM, the EPROM, and the EEPROM, Intel had to exit those industries once they became competitive.[15]

In my estimate, Intel was, on average, more than a year behind the trend line in the development of semiconductor technologies during the 1980s.[16] The reputation of the technology development groups suffered as process after process failed to perform as claimed. As it exited the DRAM business, therefore, Intel began to integrate, discipline, and standardize manufacturing and process development. Portland's technology development group stopped all work on DRAMs and was integrated with remnants of the microprocessor process development group from California. This became the seed for Intel's technology integration capability, gradually developed under the broader umbrella of the technology development organization. The first project of the new group was a shrink of the 386 chip.

Next, work commenced on process technology for the 486 chip, the first microprocessor project driven start-to-finish by the reborn technology development group. The group collaborated with former logic technology developers in a DRAM factory now dedicated to microprocessor process development. All technology choices from this point on would be optimized for their target product system, the microprocessor. The context of technology development efforts had therefore become much more narrowly and precisely defined.

Intel was changing the core foundations of its process development efforts. In the words of the vice president of technology development and manufacturing, Intel witnessed "a massive transition towards much more data-driven decisions. [This involved] a massive education program, [with] everyone put through SPC training in 6 months. [The current process] complexity requires much more discipline." Moreover, the roles of the different organizations were more precisely defined. Research efforts were clearly aimed at generating process technology options, such as new materials or production approaches. The evaluation and selection of all process technology possibilities (provided by internal researchers or by suppliers) was made in a unified, focused way by a single, experienced group of individuals within the technology development organization. This evaluation and selection was made based on precise targets derived from cost and performance objectives. As noted by the vice president of technology development and manufacturing, "The issue is to establish targets for each piece, and then consider what technology can do this."[17] Once the selection was finalized, any change in equipment or process characteristic would need approval from a centralized committee.

Intel's process developers also began to rely on closer ties with equipment suppliers for knowledge and development work. The vice president of technology development and manufacturing explained the shift in this way:

> We used to lag DRAM manufacturers. We could thus go to [a supplier] and ask, 'What are the DRAM manufacturers buying? We want it.' Now we are pushing ahead of DRAMs, or at least catching up. . . . The cross-over point has begun to take place in the last couple of years. . . . We can no longer simply buy off-the-shelf technology that was developed for DRAM. Equipment suppliers have now become the bottleneck. We have five people now at [one critical supplier] helping them to get up to speed. [At this point] we probably know more about their suppliers than they do.[18]
>
> This industry has matured to the point that you are building something as complex as a 747. It is a massive program management challenge. . . . The biggest transition was to get [suppliers] to stop doing things by seat-of-pants. We now want to see data, use statistics, SPC, etc. . . . Discipline is now the key.[19]

Intel became much more demanding of its suppliers: "Equipment is [now] selected subject to strict parameters [regarding] process capability and extendability . . . equipment suppliers now must develop [prototypes of the] equipment years in advance so that it is running by the time Intel needs it. [Our] process choice can be driven by what equipment is available."[20]

By 1991, the changes at Intel had become a massive effort to discipline process development and ensure a timely production ramp. For the 0.5 micron process generation, for which the first product was a 486 shrink, a manufacturing-scale development facility (D2) was built in Santa Clara. Technologists from process integration in the technology development group conducted equipment characterization and ran test wafers in the manufacturing environment as it evolved. In addition to more fully simulating a manufacturing fab, the large scale of the development fab allowed an unprecedented amount of experimentation. Whereas in the past some Intel process development groups would require as few as five good wafer batches[21] before releasing the process to manufacturing, now the development group alone produced test batches at capacities of as many as two thousand wafers per week. This was more than twice the capacity of development fabs of most comparable American firms at the time, and more than four times the experimental capacity of typical Japanese semiconductor manufacturers. Literally millions of test wafers were run before the process was qualified for manufacturing. This development effort was explained by a senior project member:

> What matters is first choosing the right process monitors. These are a set of variables that give us a solid, quantitative indication of product quality, including performance, reliability, and yield. We then have a focus on running a large number of test vehicles, changing the process until the monitors indicate that the right objectives are met. It is essential to run enough trials, both to experiment with different approaches, and to obtain statistically significant results.[22]

Process development was complete only when all technical options under consideration had been fully explored; when all process technologies and equipment specifications had been fully characterized, specified, and documented; and when high manufacturing yields had been reached and sustained on the development line. Ultimately, the development fab itself was transitioned to full-scale manufacturing.

Next, Intel built 0.5 micron capacity in additional fabs. The newly proven technology was more than transferred; rather, it was meticulously copied from the Santa Clara development fab. Identical equipment was implemented—and in a manner strictly consistent with guidelines set by the developers. An effort was made to copy every detail of the manufacturing process. Differences were explicitly documented and nothing was taken for granted. Managers refer to this philosophy of technology transfer as "Copy Exactly." Once manufacturing commenced, proposed changes required approval from a central technology development/manufacturing committee, and would be implemented across all fabs. Today, managers joke that even the height of process technicians must be identical in all fabs.

As many Intel managers willingly admit, the new process does slow down the pace of incremental improvements made in the manufacturing fabs. One senior manager in technology development explained the difficulties they now encounter. "The improvement process is very slow—must get [all relevant] production sites involved, to some degree, in deciding to agree to a change. There is no overall boss who orders new procedures. There are few incentives to change and improve. . . . All this means that the process must be much better when it goes into production." The slower improvement rates in the factory, however, are more than made up for by faster development cycles and reduced risk. Another manager elaborated. "In the past, yields were so low that virtually any change [in the production factory] would lead to an improvement. This means that risk was very low. Now, yields are much higher, the consequence of error much greater, and the risk of change is much greater."[23]

In successive process generations, Intel has continued to pursue a focused organizational approach to making technology choices, coupled with intensive experimentation for the development of new process technology. Technology choices are coherently made by a process integration group that is part of the technology development organization. By 1995, Intel had significantly increased its rate of new process implementation—the speed increase measured at more than one year, just between 1992 and 1995. They also utterly eliminated the U-shaped yield curves, achieving seamless transfers from development to manufacturing and from fab to fab. These factors have entrenched Intel as the leader in the microprocessor industry.

Characterizing the U.S. Turnaround

Intel's changes are representative of a broader pattern, visible by analyzing the evolution of technology integration in U.S. organizations during this time period. The focus here is on three firms: Intel, Texas Instruments, and IBM. The data is aggregated, since details of the individual organizations cannot be released for confidentiality reasons.

Knowledge generation through experimentation is the first critical mechanism in technology integration. At semiconductor firms, the experimental cycle is performed by running batches of test

TABLE 8-1

Changes in Knowledge Generation Through Experimentation

	1μm (1983)	0.35μm (1993)
Experimentation capacity (wafer starts per week)	358	1015
Average experimentation throughput time (weeks)	6.3	3.4
Minimum experimentation throughput time (weeks)	1.4	.8
Proportion experimentation/manufacturing equipment same	62%	88%

Four measures of experimental capability are examined. The first is the experimentation *capacity* available to each organization before committing to technological choices (in wafer starts per week). The more the capacity, the larger the number of experimentation cycles that an organization can conduct in a given period of time. The second is the *average throughput time*, which indicates the length of the mean experimental iteration. The third is the *minimum (rush) throughput time*, which indicates the shortest possible iteration. This is important since the most critical experiments (that is, the fabrication of the first functional sample) are often conducted in the shortest possible time, not in the average time. The fourth measure of experimental capability is the *representativeness of the development fab*. This is measured by the percentage of process equipment that was the same as that of the future production fab—the higher the percentage, the more the experimental trials will generate knowledge that is representative of the final production context.

wafers in large facilities (development fabs) that simulate the manufacturing production process. The more cycles conducted, the quicker and more representative they are, and the faster knowledge about technology-context interactions can be generated. Data on changes in experimentation during the period under scrutiny are summarized in Table 8-1.

The table shows a striking shift in the experimentation practice of TI, IBM, and Intel. As implied by my account of the evolution at Intel, these organizations substantially increased their experimental capacity. From an average 358 wafer starts per week in 1983, average development fab capacity increased to 1,015 in 1993. When combined with the typical time period in a project for experimentation, as well as the number of samples per wafer (which increased substantially, due to changes in wafer size), the growth in experimentation capacity translates into millions of additional experiments that could be conducted in any given project. This increase enabled a much greater range of technological options to be tested before committing to a set of choices. Further, each test could be conducted more rigorously, including a larger number of trials for increased statistical significance.

Moreover, both average and minimum throughput time declined steeply during the same period. This is connected to the increase in capacity since, all things being equal, the greater the capacity, the shorter the throughput time. The massive experimentation capability in place would be used not only to increase the average number of experiments conducted each week; some would be held in reserve to conduct large bursts of rapid experimentation when needed. Over the same period, the proportion of equipment in the development fab matching that in the manufacturing environment increased from 62 percent to 88 percent, indicating that the U.S. organizations were attempting to simulate the manufacturing environment more accurately in their integration facilities. These changes were coupled with a shift away from using a separate research fab, which was perceived as not being representative enough.

Between 1983 and 1993, this group of U.S. organizations built superior experimentation resources to support their technology-integration activities. Their goal was to create experimentation fabs that behaved as much like actual manufacturing facilities as possible, allowing the thorough investigation of a broad set of technological options. Aided by the other mechanisms described below, this significantly increased their ability to make effective technology choices.

Data analysis also revealed some changes in the extent of both the education and experience of the key personnel undertaking technology-integration tasks. To assess how the firms built their knowledge and personnel base, managers and engineers were asked to report on twenty to forty members of the teams primarily responsible for the critical technology integration tasks of problem definition, option selection, experiment design, and final technology selection. Some results of this data collection are summarized in Table 8-2, which reveals that these key employees possessed more education and more experience in 1993 than did comparably placed employees in 1983. While a good foundation had already been present in the 1980s, it was deepened during the early 1990s. In addition, the increases in knowledge generation and application allowed this education and experience to be leveraged more effectively.

Table 8-3 shows that project organization and resourcing also evolved to change the method by which this knowledge was applied to technology selection in the projects. Several trends are apparent. First, the total commitment of resources to R&D rose sharply. This

TABLE 8-2

Changes in Knowledge Retention: Experience

	1µm (1983)	0.35µm (1993)
Percent recruited specifically for semiconductor science and engineering	92	100
Percent with graduate education at recruitment	80	83
Percent with Ph.D. research experience	50	58
Percent with previous project experience in semiconductor process R&D	69	73
Percent with more than two or three generations of experience	35	54
Average tenure (years)	9	10

is not surprising, given the growing difficulty of semiconductor development through the period, and it is a phenomenon observed throughout the industry. The lion's share of this resource increase took place in the integration and development groups. The proportion of personnel resources allocated to these groups jumped from 67 percent to 91 percent—a jump due almost entirely to resources being aimed at process integration and indicating a growing role for its activities. Yet, these firms actually decreased their absolute commitment to the research phase of the project. The implication, however, is not that the overall semiconductor research expenditure decreased (for it did not), but that *the role of research as a separate project phase was de-emphasized.* Finally, the scope of the project changed, with increasing roles for integration and decreasing roles for development and manufacturing. Process equipment design was almost entirely subcontracted to suppliers. Integration group members strengthened their control over process equipment purchasing, usually at the expense of plant managers and other representatives of manufacturing. The diminishing role of manufacturing is also shown by the sharp increase in yield at transfer, which reduces the emphasis on process improvement during the ramp-up stage at the plant.

The conclusion from this data is that the focus on integration tasks and the role of process integration groups grew substantially over this period. This conclusion is backed by formal organizational

TABLE 8-3

Changes Related to the Knowledge Application Process

	1μm (1983)	0.35μm (1993)
Research Resources		
Person-years	58	30
Percent of total project	33%	9%
Integration and Development Resources		
Person-years	120	305
Percent of total project	67%	91%
Project Scope		
Lithography equipment developed in-house	67%	0%
Etching equipment developed in-house	33%	0%
Equipment purchasing controlled by integration group	0%	67%
Yield at transfer to manufacturing	27%	71%

changes. By 1993, technology integration resources were fully dedicated to this task in all three organizations (none had been so dedicated in 1983). The role of research and development, strictly defined, was de-emphasized, while the responsibility of manufacturing became almost exclusively focused on day-to-day operations rather than on process development. At the same time, the role of external suppliers was increased. In essence, by 1993, activities related to (process) technology integration constituted the bulk of a process development project in each of these organizations, ensuring that critical technical decisions would be made with a holistic perspective. This, combined with a deep experience base and the significant increases in experimentation capability described above, laid the foundations for the U.S. turnaround.

Semiconductors in 1995: Comparing U.S., Japanese, and Korean Practice

By the mid 1990s, the semiconductor industry comprised an impressive group of organizations. Virtually each competitor that remained, whether in the United States, Japan, or Korea, could bring highly sophisticated technologies to market rapidly and efficiently. Although competitors were capable of comparably effective performance, however, it is not true that they had developed identical research, integration, or development processes. In implementing their improvements, U.S. firms had not copied the Japanese but developed a distinct approach that was more suited to their own environment. Korean firms, as shown below, developed yet a third model, one that embodied some characteristics of both the U.S. and the Japanese approaches without taking either one for granted. This section describes and contrasts these approaches by drawing on data from projects conducted in the 1990s.

Different Foundations: Experience and Knowledge Retention

The foundation of the U.S. turnaround differed from that of firms competing in the Japanese or the Korean environment. Table 8-4 shows this by focusing on two critical characteristics of the resources used in projects drawn from each country. First, it shows that resources in Japanese and Korean projects counted on lifetime employment, while this concept was almost completely absent in U.S. firms. Hence, project turnover in Korea and Japan was much lower, and the emphasis on individual development and training was higher. The second critical number concerns the percentage of technology integrators that had research experience through a Ph.D. The data shows that university research experience is standard practice in the United States but quite rare in Japan. This phenomenon is linked to deep differences in the university traditions in each country. In the United States, the Ph.D. is a long-standing institution, respected and funded by external grants. In Japan, it is an extremely rare degree, generally only pursued on a part-time basis while the student is employed. The research is funded by the company and conducted essentially as part of its internal projects.

The numbers in Table 8-4 imply stark contrasts between the United States and Japan concerning the availability of external knowledge. U.S. companies have access to the rapidly evolving base of fundamental knowledge in science and engineering through universities. Moreover, they have access to the latest technologies from competitors by transfers of individuals. In Japan, these sources are highly limited, and most of the fundamental knowledge must be either developed internally or acquired from suppliers. Korea, for its part, appeared characterized by an intermediate model. Although lifetime employment was the norm once employees joined the company, many project participants had been hired from other firms, mostly in the United States. Korean firms also showed a higher frac-

TABLE 8-4

Career Dynamics and Project Experience

	U.S.	Japan	Korea
Fraction with lifetime employment	14%	100%	100%
Fraction with university research experience through Ph.D. programs	59%	7%	24%
Fraction with no previous project experience	34%	14%	14%
Fraction with one generation experience	28%	34%	22%
Fraction with two generations experience	23%	30%	23%
Fraction with more than two generations experience	15%	23%	41%

tion of Ph.D.-level hires than did Japanese firms. In some competitors, the fraction even approached 50 percent. Interestingly, many of these Ph.D.s were awarded from U.S. universities. In contrast to the Japanese group, therefore, Korean firms benefited from significant external sources of knowledge, primarily knowledge emanating from the United States. This appeared to fuel the substantial technological improvements they accomplished over the last decade.

These country differences are reflected in the project experience data, also shown in Table 8-4. The lifetime employment practices of Japanese and Korean firms lie behind their deeper levels of project experience. In the United States, the high percentage of individuals with no previous project experience reflects the common practice of implementing targeted hires from university Ph.D. programs to fill the need for expertise in specific disciplines. Examples might be Stanford for process simulation or MIT for lithography techniques. Individuals are hired from these programs with immediate tasks and objectives in specific projects.

There are also significant differences in the knowledge foundations for research, integration, and development in different parts of the world. In Japan we find an internally focused model that relies extensively on the internal development of individuals and on critical supplier contributions (or, more recently, alliances with U.S. firms). U.S. firms make use of internal research as well as hires from other firms and students from local universities to provide the latest technological concepts. Korean firms adopt a hybrid model, scanning the world for talent but internalizing it permanently through lifetime employment practices. These factors are difficult to change. They reflect deep-rooted differences in university traditions and labor market characteristics around the world.

Different Tools: Experimentation and Knowledge Generation

The different foundations for research, integration, and development observed in Japan, Korea, and the United States influence the process followed for knowledge generation in a project; this has an impact on the approach to experimentation indicated by the values of experimentation capacity and iteration time shown in Table 8-5. The deep project experience base found in Japanese and Korean firms works essentially as a substitute for experimentation. Rather than building the massive development fabs seen at Intel, for example, Japanese and Korean projects relied intensely on experience to drive technology selection. The intuition of experienced project members is then confirmed through a rapid sequence of more limited, carefully selected experiments. This approach contrasts with that found in the United States. The knowledge brought in from universities offers great potential performance improvements but must be thoroughly tested out before it can be applied to practice. The investment in experimentation is central to this, exemplified by the high average experimentation capacity value shown in Table 8-5.

Different Process: Focus and Knowledge Application

The deep differences in knowledge generation and retention approaches in the three geographical areas are reflected in additional contrasts revealed by analyzing actual processes and organizations. The most striking difference is the organization and scope of integration and development activities. The U.S. model for integration and development is exemplified by the Intel description above. Projects are characterized by a high degree of centralization, whereby a group of integration experts is charged with conceptualizing the next process generation. The team is grouped together in one location and has massive experimentation at its disposal; further, once the choices are made, the team remains as a driver, working in the development fab until the new process generation is fully operational and generates high yields. The average yield for transfer to manufacturing in the United States (58 percent) is thus markedly higher than practice in either Japan (21 percent) or Korea (25 percent). The integration and development organizations in the United

TABLE 8-5

Experimentation

	U.S.	Japan	Korea
Experimentation capacity (wafers/week)	917	457	417
Average experimental iteration time	16	13	6
Minimum experimental iteration time	5	7	5

States are thus responsible for delivering a complete, high-yield production process to manufacturing, suggestive of the extreme "Copy Exactly" philosophy at Intel.

For their part, project activities in Korea and Japan are networked over several functions and suppliers. Research and manufacturing share much more of the tasks at the project level. In contrast to the process used in the early days at Intel, however, this does *not* mean that technology choices are made in a scattered fashion. Rather, a small but unified group of technology integrators usually leads the selection process for the entire production system and is responsible for its successful implementation. While responsible, this group does not actually perform the ramp-up, but hands most tasks off to manufacturing at relatively low yields.

The significant project experience in Japan and Korea is reflected in a higher percentage of reused equipment (around 75 percent for Japan and Korea versus 60 percent for the United States). This is not only because of the technology integration group, however. Manufacturing retains considerable power over equipment selection. The R&D head in one of the leading Japanese companies confirms this:

> We try to standardize equipment between R&D and manufacturing, but if the plant manager wants to change something, he will. Why should I say anything when we know that his is the best semiconductor factory in the world?[24]

Different Outcome: Evolution and Revolution

The contrasts between U.S., Japanese, and Korean organizations are quite significant. All models achieve integration, although they do so in different ways. In synthesis, the U.S. model is reminiscent of the revolutionary project descriptions in the previous chapter, combining extensive knowledge transfers from research with massive experimentation. Project organization is highly centralized, with all relevant integration resources co-located under one roof, with access to extensive experimentation. When compared to the U.S. model, Japanese projects are instead more reminiscent of the evolutionary approach, combining extensive project experience with selective, linear experimentation. Organizationally, the Japanese projects were more distributed, gathering knowledge from a variety of functions and organizations, with extensive impact by research, manufacturing, and the supplier network. Korean projects appeared to follow a hybrid model, located somewhere between those of the United States and Japan. While there was considerable variance between Korean organizations, the tendency was to combine extensive project experience and a more careful approach to experimentation (similar to the Japanese model) with extensive research experience, primarily from universities (following the U.S. model).

Table 8-6 confirms these expectations by showing that Japanese projects achieve higher technological yield and lower technological potential than do U.S. projects, on average. Japanese projects therefore achieve evolutionary outcomes when compared with projects in the United States. Korean projects are indeed a hybrid—on average better than U.S. projects on yield but worse than U.S. projects on technological potential.

Building Technology Integration Capability

In summary, by examining the striking improvements achieved by a group of firms in the semiconductor industry, this study reveals several important points relevant to the implementation of technology integration in practice. First, organizations can change their technology integration process and do so substantially and in relatively limited periods of time. Additionally, the changes are actually associated with improvements in project performance. Effective firms did not change by simply copying successful competitors, however, but approached the challenge of technology integration by adjusting to the strengths and weaknesses of their local environment. This, in turn, appeared to shape the evolutionary or revolutionary nature of their projects.

The discussion focused on the quintessential science-based industry; semiconductor components have traditionally been dominated by large U.S. science-based firms like IBM and AT&T. The industry has always been characterized by a close link to science,

TABLE 8-6

Average Project Outcome

	U.S.	Japan	Korea
Integrated circuit density (normalized residual)	1.01	0.74	–0.09
Technological potential (normalized residual)	1.02	0.54	–0.17
Technological yield (normalized residual)	–0.08	0.30	0.10

All numbers compare values to the industry trend. A value of *1* implies a standard deviation above the trend, while a *–1* implies a standard deviation below the trend.

represented principally by the physics and materials science academic communities. The changes described in this chapter illuminate how the interaction between traditional science and the application context has fundamentally shifted. The same firms that invented the concept of the R&D laboratory were forced to restructure it, under pressure from foreign competition.

The next chapter covers the evolution of R&D in personal computer software and workstations. These segments of the computer industry are much more recent, and the dominant firms never had time to develop the traditional R&D model described in Chapter 2. The environments are still driven by performance, however, and harnessing the potential of science—whether it is work on the efficiency of parallel computer algorithms or on computer simulation—is still essential to competition. Rather than restructure an existing organization, these firms had to learn how to build integration capability from scratch, linking the science base into an environment characterized by weekly pressure to ship products, rapidly changing user needs, and exploding network infrastructures like the Internet.

Notes

1. The data for non-Japanese firms were aggregated in the figure for simplicity. This essentially combines U.S. and Korean manufacturers. Plotting U.S. and Korean averages separately does not change the basic picture.
2. Robert A. Burgelman, "Fading Memories: A Process Theory of Strategic Business Exit in Dynamic Environments," *Administrative Sciences Quarterly* 39, no. 1 (1994):24–56.
3. Ibid.
4. G. W. Cogan and Robert A. Burgelman, "Intel Corporation (A): The DRAM Decision," Case PS–BP–256 (California: Stanford Business School, 1990); B. K.

Graham and Robert A. Burgelman, "Intel Corporation (B): Implementing the DRAM Decision," Case PS–BP–256B (California: Stanford Business School, 1991); and G. W. Cogan and Robert A. Burgelman, "Intel Corporation (C): Strategy for the 1990s," Case PS–BP–256C (California: Stanford Business School, 1991). This section combines insight gained from these studies with my own extensive field research performed at Intel Corporation from 1990 through 1995.

5. Cogan and Burgelman, "Intel Corporation (A)," 9.
6. Interview conducted by author during field research from 1990 through 1995.
7. Ibid.
8. Ibid.
9. Ibid.
10. Interview conducted by author during field research from 1990 through 1995, and proprietary Intel documents.
11. Interview conducted by author during field research from 1990 through 1995.
12. Cogan and Burgelman, "Intel Corporation (A)," 15.
13. For a detailed description of the problems, see Cogan and Burgelman, "Intel Corporation (A)," 15.
14. This calculation is based on my DRAM product introduction database.
15. See Cogan and Burgelman, "Intel Corporation (A)," and Graham and Burgelman, "Intel Corporation (B)," for additional detail.
16. See Table 9 in chapter 4.
17. Interview with vice president of technology development and manufacturing conducted by author during field research from 1990 through 1995.
18. Ibid.
19. Ibid.
20. Interview conducted by author during field research from 1990 through 1995.
21. Graham and Burgelman, "Intel Corporation (B)," 3.
22. Interview conducted by author during field research from 1990 through 1995.
23. Ibid.
24. Ibid.

Integrating Technology and Market Streams

chapter 9

IN 1993, MOST COMPUTER industry competitors were betting that video on demand (VOD) would take over the entertainment environment. Microsoft had an enormous share of the personal computer software market but was "not in the on-line business," according to some of its top executives. Yahoo! and Netscape did not exist. Less than three years later, VOD was dead, and "everyone," from IBM to the local gift shop to ordinary people, had a Web site on the Internet, making Netscape (which employed a thousand people and whose browser held some 80 percent of a market that had been scarcely visible in 1993) a household name; Yahoo! was worth a billion dollars. In a stunning about-face, Microsoft had radically changed its strategic orientation to focus on the on-line environment. To get some understanding of the speed of this change, consider that it takes about three years for a major software operating system release to move from conceptualization to market.

Moreover, no one expects this hectic pace to slow, as a range of new possibilities already looms. Would the information appliance concept challenge existing hardware and software platforms? What would drive object-oriented database standards on the Internet? Would Intel support the OpenGL graphics language? Would Microsoft NT? Such questions (and the list seemingly lengthens by the day) challenge any major project in workstations or software conducted during the mid 1990s. Clearly, these efforts must adapt to constant change in markets and customers.

New technology still drives competition, however. Workstations are still bought based on performance, and performance comes from new technology. Internet search programs function because of their search engines' speed, and that speed depends on highly sophisticated software. Developing this type of technology takes time; it takes relationships with universities, long-term commitments, and substantial expenditures. How can companies make such long-term commitments when customer preferences are continually evolving? Novel ideas are born almost weekly, arising from market turbulence or the recognition of the new possibilities arising from a still immature technical base. The complexity of the existing infrastructure and market is already considerable, with a typical Internet program already resembling the size of a PC software application.

In sum, projects confront completely different time scales. A sophisticated computer or a complex software program takes years to develop; the market, on the other hand, can shift in a matter of days. The traditional model of R&D as a linear chain of phases comprising careful research translated into a good product concept, rigorously frozen at concept approval, and meticulously executed by a highly focused development organization, simply cannot work. There is too much change.

This chapter shows that technology integration is central to innovation in the personal computer software and Internet environments. It is difficult to think of two environments more different than are Intel and Yahoo! Still, they share a fundamental need: Managing the interaction between knowledge of new, novel technological possibilities and knowledge of their complex application context. In semiconductors the novelty is primarily driven by changes in the technical base. In Internet software, much of the novelty is instead external, associated with the market. In semiconductors, the complexity of the application context is driven by the manufacturing setup. In Internet software, this complexity is driven by the network architecture and, especially, by the user environment. Therefore, while semiconductor process development is internally driven, aimed at optimizing technology to conquer manufacturing challenges, effective software development on the Internet, as this chapter will show, is externally driven, emphasizing responsiveness to the turbulent market environment.

Though the sources of novelty and complexity are different, however, the need for managing the interaction between technology and application context is, if anything, even more critical in software. In Chapter 7, I argued that the process linking technological options to application context will drive the effectiveness with which an organization can respond to threats and opportunities. While in semiconductors, major responses may be needed once every few years, in the Internet environment they are necessary many times within a single project. An effective process for technology integration is critical because it enables the joint optimization of technology and context as the project evolves. It is not, therefore, surprising that the types of knowledge generation, retention, and application mechanisms discussed in this book are critical to managing the turbulence in the new software industry.

This chapter investigates how firms in the personal computer software and Internet service industries are building the technology integration capability needed to compete; for these firms, technology integration is a central process for a model of competition based on reacting to the needs of their technological and market environment through projects. After looking at the evolution of product development at Microsoft, one of the first firms to really work at conquering the combined challenges of novelty and complexity in a systematic fashion, we turn to the processes at Netscape, Yahoo!, and NetDynamics, among the leaders in the new Internet environment. We revisit Microsoft to update its development processes,[1] and then examine critical patterns across these firms.

Microsoft's Evolution

In the Beginning

Microsoft's development process in the early 1980s was informal, with little emphasis on schedule or process. Projects were driven by a few brilliant star developers,[2] who had almost total control of a product's design. Few colleges or universities offered degrees in computer science, computers were relatively scarce, and finding programmers was difficult; like Bill Gates, most early Microsoft developers had little or no formal training in computers. Many had expertise in other fields (particularly in math and science) but had fallen in love with programming along the way. Perhaps because of their lack of formal training, most of these stars (called technical leads) did not follow the highly structured software development methodologies created by the Department of Defense and large corporate MIS departments. Their exploits were legendary:

> There was one guy who could type at eighty words a minute. That's pretty impressive, but what's really impressive was that he actually wrote code at that speed. He'd write a 10,000 line application in two days, then if it didn't work, he'd throw it out and start again from scratch. He'd go through this process two or three times, until he ended up with a working program. Not only did it take him less time to do this than if he had sat down and tried to think everything out in advance, but the program that resulted was better too. Because he had implemented the same program several times before, he knew how to avoid all the pitfalls, so his code became very clean.[3]

One of Microsoft's early developers described his ideal audience:

> I design user interfaces to please an audience of one. I write it for me. If I'm happy I know some cool people will like it. Designing user interfaces by committee does not work very well: They need to be coherent. As for schedules, I'm not interested in schedules; did anyone care when War and Peace came out?[4]

Not surprisingly, given such approaches and attitudes, what the product would look like and when it would be released frequently remained mysterious. Bill Gates was one of the few controls in the process, influencing developers primarily during the intense project review meetings he attended. Meanwhile, the company's image was that of a firm producing technically excellent but difficult-to-use products. At the same time, software applications became significantly more complex, challenging project teams even further. The company's first program manager recalled this pattern.

> Development of outstanding software was starting to require much greater attention to detail than to sheer performance. What mattered most was attention to how the individual features fit together into a well-designed, coherent product that was attractive, reliable, and fun to use.[5]

Increases in the software market's complexity led to corresponding modifications in product development. No longer could one technical lead manage the entire undertaking, and the program management function was created to share the responsibility for decision making. The first program manager was assigned to work on the design of a spreadsheet for the Apple Macintosh in 1984. As he noted, he "became a sort of service organization for the development group, . . . I helped document the specifications, do the manual reviews, and decide what bug fixes were important and what could be postponed to a later release."[6]

By the late 1980s, several people shared leadership for the development of a new product. The project lead was responsible for overseeing the product development effort, including handing out programming task assignments, scheduling, and coordinating the development effort. The technical lead had final say in all technical decisions, code reviews, and programming standards. The product manager handled all the marketing issues such as competitive analysis, positioning, packaging, and advertising. The program manager's job was to integrate and coordinate the efforts of everyone involved in the project. Program managers were also directly responsible for product concept and specification. Finally, on-line and print-based leads handled the user education functions, and the localization lead oversaw customizing the program for various international markets.

Microsoft's software development projects grew substantially in size and complexity. In the mid 1980s, a project aimed at a major application might include five developers; ten years later more than fifty might be needed. The enormous challenges that this posed for developers are illustrated by the development of Word for Windows 1.0 (shipped in November 1989); it was a pathbreaking experience for the company and prompted a number of critical changes in Microsoft's process.

Word for Windows 1.0

Microsoft introduced its first high-end word processor for the PC, PC Word, in late 1983 to lukewarm reviews and mediocre sales. In September 1984, Gates, convinced that a key opportunity was being squandered, urged the development of a new, revolutionary word processor that would run on the Windows operating system (then in development) and exhibit some extremely innovative features that would make Microsoft the leader in the field. He made three experienced people responsible for the project, code-named Cashmere. The program manager subsequently assigned to Cashmere recalled the project:

> Gates put three real hotshots on the project. The first had single-handedly written the first version of PC Word and became the project

> manager. The second had a Ph.D. in psychology and was responsible for user interface and documentation. The third had been at Wang and was supposed to know the word processor business inside and out.[7]

Gates directed the team to "develop the best word processor ever" and finish the project as soon as possible. October 1985 was the scheduled completion date.

The Cashmere team quickly created a concept of the application, communicating it to Gates via papers and specification documentation. Way ahead of its time, the idea was to integrate a uniform user interface and data structures for multiple purposes at the lowest level. That is, they planned to structure the word processing program and data so that it integrated seamlessly into other types of applications, such as spreadsheets and databases, thereby blurring the distinction among product categories. Some features to be included were electronic mail, document protection, facilities to build mailing lists, and primitive spreadsheet capabilities. In other words, team members were planning to develop something that resembled Microsoft's entire Office Suite, released in the mid 1990s. All in less than a year!

Needless to say, these objectives completely overstretched the team's capabilities, and a year passed; several software developers were brought on to prototype software, but their work stalled. By the beginning of 1986, the scheduled ship date was still a year away, and Gates began to pressure the team to produce visible results. Eventually, several team members left the project, and when replacing them Gates decided to use the program manager concept, which was then being implemented elsewhere in the company. Three Microsoft veterans were brought in: Along with the addition of a program manager, the PC Word development lead took over the same role for Cashmere, while a well-regarded developer assumed the technical lead role. A new marketing manager was also added. A new team member described the scene:

> We all thought that the project was much farther along. We ended up throwing everything that had been done out and started from the code used in Word for the Macintosh. We were a year behind schedule from the first day we started.[8]

The project was renamed Opus, and a new team of developers was formed, staffed almost entirely by new hires. Very few developers had had experience with other Microsoft projects.

The second half of 1986 and the first half of 1987 were dedicated to writing a new specification for the product while pressures mounted, once again, for visible results. The schedule continued to slip into 1988, and the pressure increased to a disturbing level. A software development engineer recalled this period:

> We were under a lot of schedule pressure. Some managers seemed to regard the schedule as a contract between them and the developers. Furthermore, when development came up with a new schedule, management questioned every estimate.[9]

Upper management kept up the pressure, the results of which were noted by a team member:

> Opus got into a mode that I call "Infinite Defects." When you put a lot of schedule pressure on developers, they tend to do the minimum amount of work necessary on a feature. When it works well enough to demonstrate, they consider it done and the feature is checked off on the schedule. The inevitable bugs months later are seen as unrelated. Even worse, by the time the bugs are discovered, the developers can't remember their code so it takes them a lot longer to fix the problem. Furthermore, the schedule is thrown off because that feature was supposed to be finished. These problems aren't unique to Microsoft—every company in the industry faces them.[10]

In April 1988, the development lead had to take a medical leave of absence and was replaced by a programmer with less than two years of experience.

Over the next several months all the required features were coded (though not debugged), and in October the team declared that the Code Complete milestone had been reached. Code Complete meant that all that remained to be done was to debug and optimize the code for performance. This phase was called *stabilization*, and once the code had stabilized (all known bugs were fixed and performance was adequate), the product could be shipped. For scheduling, Microsoft used a rule of thumb that the stabilization phase typically lasted three months.

It soon became clear that Opus was not likely to follow the three-month rule. Although developers were fixing bugs quickly, the testers seemed to find new ones just as fast. By the spring of 1989 the number of active bugs remained relatively constant, but, during the summer, an initiative emphasizing the quality of changes rather than the quantity of changes was instituted. For the first time, testing people were invited to development code reviews, and ownership of the code was stressed. By late fall, the program had stabilized and Word for Windows version 1.0 was released on November 30, 1989. Naturally, the product when it shipped was nothing like the original specification put together by the Cashmere team. Its architecture, features, and performance had been scaled down considerably to make the project feasible. Still, the word processor was viewed quite favorably by both critics and customers, and it built to an enormous business (more than $500 million) in its first years.

Technology Integration Failures

As is evident from the preceding description, the development of Word for Windows 1.0 was fraught with problems. The product shipped four years after its original target date, causing Microsoft's Windows operating system (shipped in 1985) to be on the market for years without a useful word processor application. Windows did not penetrate the market until 1990. Moreover, the process was extremely unpredictable; the ship date was in serious question even six months before project end. Finally, Cashmere/Opus completely drained its members, forcing several either to take leaves of absence or to quit the company altogether.

Many of these problems can be traced to the earliest project stages. When the first team put together its original, impossible specification, it set incredibly high expectations for all subsequent efforts. Full of novel elements, these objectives were submitted to Gates without any substantial effort at experimentation; clearly the team members' guess about the product's feasibility was very wrong. The rest of the project was rife with efforts to change the significant inertia that had built up around it. As the "infinite defects" description shows, these efforts often failed, and even in later project phases programmers felt under pressure to add features at an impossible rate, at the expense of testing and reliability.

Winword thus failed to incorporate many of the basic elements of the technology integration framework presented in this book. Although it had revolutionary goals, the project did not include enough experimentation, particularly before committing to its most critical choices in features and technology. Given frequent turnovers in team membership, the project suffered from the inexperience of its critical decision makers. And responsibility for technology choices and design was uncertain, as the role of program manager was confused with that of the many other leads in the project. As such, Winword suffered problems in all three mechanisms underlying effective technology integration: knowledge generation, retention, and application.

Nevertheless, as time progressed, a better process for experimentation and testing was gradually established, with better collaboration between development and testing organizations. And as it gained in importance, the program manager function assumed a more active role in managing the evolution of the specification. Even the experience level of the team members increased, naturally, as the project struggled on. These changes eventually made the specification converge with reality, and a good, reliable product was finally shipped.

Microsoft's New Process

Partly in response to the problems with Cashmere/Opus, in 1990 several proposals were brought forth to improve Microsoft's development process. These were aimed at improving the reliability and quality of product development while retaining its flexibility and responsiveness. Microsoft managers grappled with two different possibilities.

One approach was to try to do what Cashmere/Opus failed to accomplish—hold developers to the original specification. Some people argued that if enough time were spent writing the original specification, and enough discipline existed in the team to respect that specification throughout the project, the process could be made to work quite effectively. The second approach was detailed in a now-famous "zero defects memo" written in 1990. The idea was to let the specification change during the course of the project—the uncertainties in technology and market were just too great to be able to predict everything up front—but to do so in a controlled fashion, driven by daily integration and experimentation efforts.

Microsoft's senior management opted for a new development process for applications based on the second approach. The process worked as described below (see Figure 9-1; also see Cusumano and Selby, 1995, for much additional detail). After conducting considerable market analysis and several focus groups, product managers described the unfilled needs that the new product should address. The program manager then translated these general market needs into broad project objectives contained in a vision statement and outline product specification. Developers worked from the outline specification to bring the objectives to life. As developers strove to code the desired features and to optimize software performance,

FIGURE 9-1

A Flexible Approach to Software Development

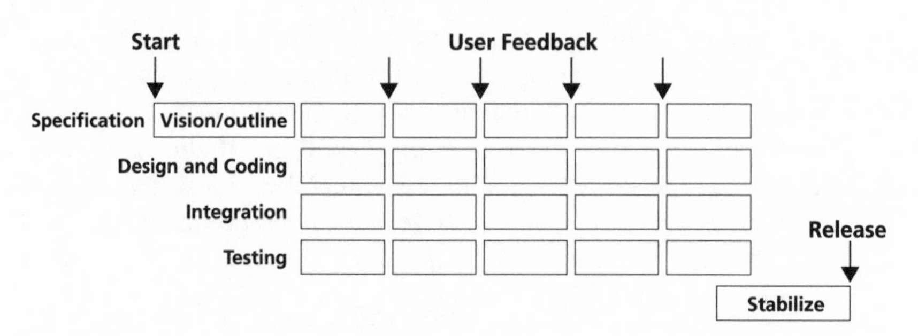

Source: Alan MacCormack and Marco Iansiti, "Living on Internet Time: Product Development at Netscape, Yahoo!, NetDynamics, and Microsoft," Case 9-697-052 (Boston: Harvard Business School, 1996).

they typically discovered problems and possible improvements. These were discussed with program managers (each major project was now staffed with several) who, if necessary, changed the specification. This cycle was repeated frequently until the project was finished.

The new process also allowed for much deeper and more systematic customer feedback. Each major prototype release would be tested with customers in individual usability sessions or focus groups, thereby obtaining market feedback as the project evolved. The approach therefore allowed the project to remain responsive to market changes by adding or removing features in a controlled fashion.

In projects involving more predictable architecture, the features in the outline specification would be divided into bundles, each to be coded and tested sequentially. The product would thus progress from a simple (but functional) skeleton to a complete application in a controlled and testable fashion that program managers could closely follow. Programming tasks were split among the many team members and integrated as frequently as possible (often daily) into a functional software prototype of the entire system. The ideal was to try to have "a shippable product every day,"[11] adding desired features and product improvements in order of their priority. Developers worked on the most critical features first and gradually expanded the product's feature set until it was time to ship the product.

Organizational and cultural changes complemented the new process. Microsoft program managers became the primary drivers in a typical development project and strong influences on the company's future. One program manager described the role:

> The key is to be able to create and articulate a vision of the product. The ideal program manager probably would have a development background. The key component, however, is to have the knowledge and ability to talk to developers in their own language. They need to be respected by development. A program manager also needs strong design skills. Finally, he/she also needs good people management skills. In practice these are often lacking in our program managers.[12]

Another program manager elaborated:

> A good program manager must be comfortable with the technical aspects of the specification and know how to change it. The specification gradually evolves until the code is complete. It is important that tradeoffs are clearly presented during its evolution. These must be clearly communicated to the team.[13]

Program managers at Microsoft had thus become the de facto technology integrators in the project; they had responsibility over the entire specification and were charged with making the most critical decisions and tradeoffs. They had to keep a holistic perspective and make sure that the technologies developed by the coders worked together to create a well-integrated, reliable product.

The new process, therefore, moved Microsoft toward more systematic approaches for knowledge generation, retention, and integration. By working intensively on methodologies for experimentation and testing and on the role and experience of its program managers, the new process allows a good degree of control and predictability while retaining the flexibility to respond to market changes.

Shooting the Rapids: Technology Integration and Flexibility

With its new development process, Microsoft began to tackle a problem fundamental to all of its competitors. The current software and Internet service environment is characterized by extraordinary levels of technical and market novelty, creating a virtually unprecedented need for flexibility and responsiveness. The challenge is to allow the appropriate level of responsiveness while keeping the development process under control. The consequence is a new set of imperatives for managing the product development process that sharply contrasts to established models.

Many existing models for development emphasize the need for avoiding unnecessary change and uncertainty in the evolution of technology and market needs. Their focus is on a structured process with clearly defined and sequential phases through which the future product is defined, designed, transferred to the manufacturing plant, and rolled out to the market. Performance is related to mechanisms that add clarity and stability to the project, such as a clear project definition phase and a stable product concept and specification. The goal is focused and efficient project execution, which entails strong project leadership, simultaneous engineering, and team-based organizational structures.

In such models, a distinct separation exists between concept development and implementation. A good project is characterized by extensive and focused activities that identify customer needs and assess technological feasibility, followed by the development of a detailed and thorough concept document, which is then presented for approval. If approved, the concept is frozen and attention shifts to implementation. And a good project is characterized by minimal changes after concept approval: If the earlier work has been done "right," later changes, which are inherently expensive, should not be necessary.

This model works well when technology, product features, and competitive requirements are predictable—that is, in environments in which novelty is low. If changes in technology or market during the project's life can be forecast, delaying the concept-freeze milestone provides no benefit; the aim of the entire effort should be on completing all stages as quickly as possible. Product development speed will influence the ability to react to competitors and has even been linked to resource utilization.[14]

By contrast, the models adopted by Microsoft and recently perfected by several other firms in the software and Internet environments embrace change, rather than fight it; these models exhibit a development process characterized by flexibility. In Microsoft's development process (see Figure 9-1), the concept development stage continues as long as the specification evolves. This occurs until the brief final stabilization stage, thereby allowing the project to respond to internally detected problems and customer feedback as well as to changes in the market environment. The work done on the project is tested and improvements are included in the next specification cycle. This approach is similar to that used by Silicon

Graphics, as described in Chapters 3 and 6, for workstations. The key to the process is in the ability to gather and rapidly respond to new knowledge about technology and application context *as a project evolves.* This is an approach I call "shooting the rapids" (Iansiti, 1995e), since it involves starting on the product development effort without a precise idea of how the effort will end; it is a flexible approach that ensures that the project team does not crash into unseen "rocks" of novelty and complexity and sink.

Significant, systemic changes in a project's definition and basic direction are managed proactively by creating a development process and a product architecture that increases the speed with which the organization can react to changes. Much new technical and market information will emerge during the typical time line of a project. In "shooting the rapids," the emphasis is on creating the ability to respond to newly discovered information during the project itself. In Figure 9-2, this translates into the ability to move the concept-freeze milestone as close to market introduction as possible.

In a highly novel and complex environment, speed is a subtle notion. Total lead time is clearly important, since it indicates the total time taken to fulfill initial project objectives. The concept lead time and development lead time measures defined in the figure are critical measures in their own right, however. The concept lead time is the window of opportunity for including new information and for optimizing the match between technology and system. (See Tyre and Orlikowski, 1994; and Iansiti, 1995e.) Conversely, development lead time indicates the time during which the window is closed, the product's architecture is frozen, and the project is unable to react to new information. Although the total lead time is the same in both cases shown in Figure 9-2, I would argue that, for novel environments, the project described in (b) is preferable to that in (a). The shorter the development lead time, the greater the ability of the organization to respond to change. The most critical measure, time to market, is thus represented by the time elapsed after concept freeze.

The model in Fig. 9-2(b) indicates that concept development and implementation are a tightly linked set of activities, not sequential phases. To help shrink the development lead time, the product's concept and detailed design are being developed simultaneously. The project navigates up and down the hierarchy of design decisions (see Marple, 1961; and Clark, 1985) iterating between fundamental, architectural choices (core concepts) and details. Although some up-front detailed design work will inevitably be wasted, the

FIGURE 9-2

Two Models of Effective Product Development

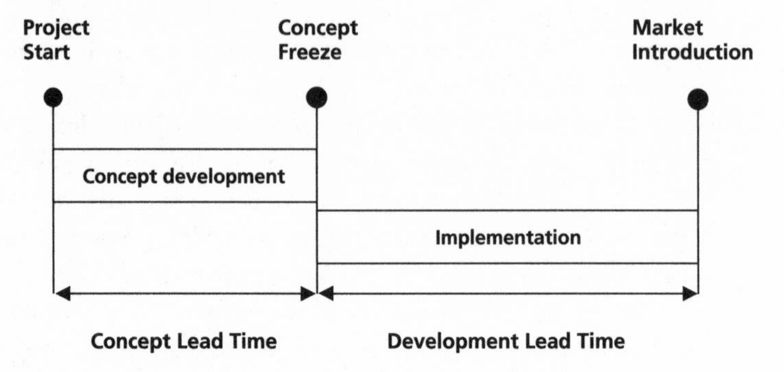

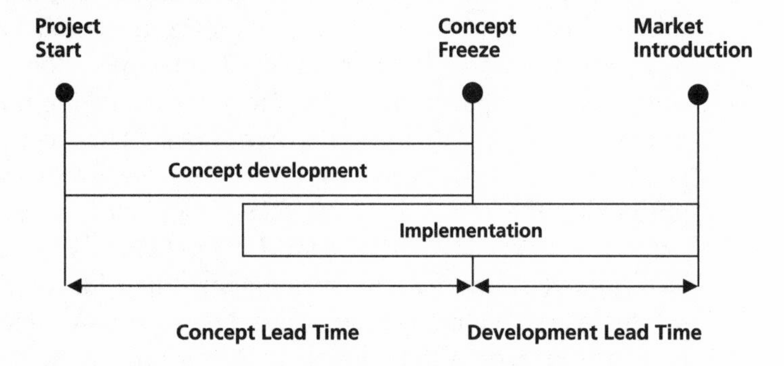

need to respond rapidly to unpredictable changes in technical or market environment makes iteration essential.

It is important to emphasize here that the flexible approach outlined in Figure 9-2(b) is quite different from concurrent engineering. Concurrent engineering models do *not* normally imply the simultaneous execution of conceptualization and implementation, but the joint participation of different functional groups in the execution of these separate and sequential sets of activities. Cooper (1983, 1990), for example, divides the development process into seven sequential stages: idea, preliminary assessment, concept, development, testing, trial, and launch; the first three are aimed at conceptualization, the last four at implementation. Clausing (1994), another champion of concurrent engineering, similarly divides the development process into four basic sequential phases: concept, design, preparation for production, and production. The flexible model presented here is different from concurrent engineering or integrated problem solving (Clark and Fujimoto, 1991). While concurrent engineering fosters the joint resolution of different functional tasks in each stage of a project, the flexible model actually implies overlapping the stages: managing the joint evolution of system and technology, of a product's concept and detailed design.

Technology integration capability is central to meeting the challenges of such unpredictable change. Product development flexibility is rooted in the ability to manage the joint evolution of technology and application context. This is in turn founded on a process that captures rich, system knowledge of product, user, and production environment and deploys it proactively to guide technology choices and concept decisions. A process that manages the interaction between technology and context enables the application of a different set of design principles, which avoids hierarchical, sequential, and rigidly defined phases. The process emphasizes, instead, rapid and flexible iterations through system specification, detailed component design, and system testing phases. This is exactly what technology integration is about. In semiconductors the iteration between technology and context was critical because it drove the optimization of critical technology choices and the rate of product performance improvement. In that business environment, technology integration capability has the added benefit of increasing the project's responsiveness by tightening the linkage between technical choice and application context.

The above argument fits with the evolution of Microsoft's development process described above. In the Internet world, the need for responsiveness is even greater. As shown in the following examples, organizations that effectively compete in this environment have developed even more extreme versions of this flexible approach. Yet the basis for their effectiveness can still be linked to the mechanisms described previously in semiconductors, mainframes, and workstations: knowledge application, generation, and retention.

Netscape

In mid 1996, Netscape Communications produced the most popular browser software for the Internet, Netscape Navigator, which commanded the lion's share of the worldwide market. It also developed and marketed a whole range of Internet-related products, such as corporate Intranet software, development tools, and software for managing Internet servers. Netscape was founded on April 4, 1994, by Jim Clark and several University of Illinois students who had developed the first graphical browser for the Internet, NCSA Mosaic. Netscape improved on Mosaic to launch its Navigator application, which rapidly became the standard. The company went public in August 1995, and quickly achieved a market value exceeding $5 billion. By June 1996, the firm had hired more than 1,100 people and was recruiting at a pace of forty new employees per week.

Netscape's development process is characterized by the very early release of a product to interested users, long before its features are established. This is done through multiple beta product versions, which are gradually and systematically improved. In this way, the customer base helps evolve the product until it is robust and complete enough for general release. A look at how Navigator 3.0 was developed will clarify the company's development process.

The Navigator 3.0 Project

After Navigator 2.0 shipped in January 1996, there were so many ideas about what should be incorporated into the next version, known as Atlas, that the development group defined some basic project objectives and soon began a software prototype. The engineering team assigned to Atlas comprised about twenty engineers focused on Navigator's overall architecture and user interface. Other development groups would work on designing program components (plug-ins), such as security routines or the capability to script applications for Sun's Java programming language. Client engineering served as the technology integrator for the product, working with elements provided largely from other internal and external groups. The team also included staff from functions outside engineering, such as marketing; in the absence of a formal program management function, the project was driven by engineering team members.

The first system-level prototype was produced extremely quickly. By February 14, a Beta 0 version was put up on the internal project Web site for use by development staff. This prototype was, naturally, incomplete, but it already embodied a fully functional browser system. On February 22, this was updated with a Beta 1 version, again for internal development staff only. The first public release, Beta 2, appeared on Netscape's Web site in early March. Additional public releases were thereafter introduced every two or three weeks until the official release date in August.

This sequence of beta versions is extremely useful to Netscape, as it enables the project to react to feedback from users and to changes in the marketplace. Beta users tend to be more sophisticated than Netscape's broader customer base and hence represent a source of valuable feedback. Most useful among these beta testers are developers from other Internet software companies, who tend to be Netscape's most vocal customers, especially those using Navigator as part of the environment in which their own products operate. Other important feedback comes from newsgroups, in which users discuss the latest software releases and log bugs in the software. The importance of such input is illustrated by the fact that the newsgroups alone were responsible for more than one hundred changes in the product's design.

During the development cycle, the team also paid careful attention to competitive products. As the largest and most powerful software developer in the industry, Microsoft is a serious threat to Netscape's position. Hence the latest versions of Gates's competing product, Microsoft Explorer, are continually monitored to compare features and layouts. Based on the feedback from these trials, the Netscape team often adds format or feature changes to the current beta version of their own software.

FIGURE 9-3

Development of Navigator 3.0

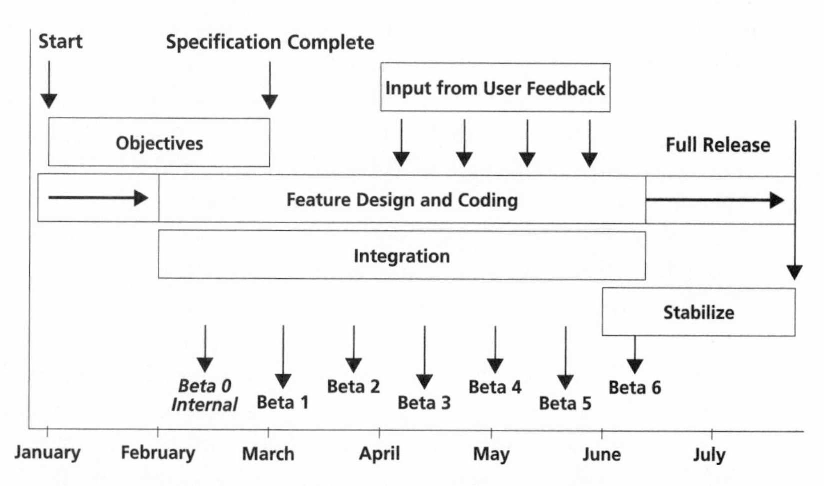

SOURCE: Alan MacCormack and Marco Iansiti, "Living on Internet Time: Product Development at Netscape, Yahoo!, NetDynamics, and Microsoft," Case 9-697-052 (Boston: Harvard Business School, 1996).

An internal project home page on Netscape's Intranet facilitates the team's communication. This Web site contains the development schedule and the product specification, which are updated as target schedule dates change or new features are added. The page also allows access to bulletin boards, on which are discussed various parts of the design, and progress charts, which track when specific features are completed and log problems (bugs) in the existing code. These management tools are especially valuable when the product moves to the beta testing phase, as the amount of information to be received, classified, and processed is significant as the product evolves. Bug charts are particularly helpful in determining when the existing beta version is thought stable enough to be released as a full version.

Yahoo!

Yahoo!, an Internet service provider, operates search, directory, and programming services for navigating the increasingly complex environment of the World Wide Web. Founded in April 1995, the company went public one year later with a market valuation exceeding $500 million. In mid 1996, Yahoo! employed a staff of eighty, around half of whom were surfers: Their job was to travel the net looking for new sites, classifying them for input into a database that was accessed by the search engine. The remainder focused on service development, advertising, and marketing.

Like many Internet-related firms, Yahoo! makes all its money from advertising revenues. For a potential advertiser, the attraction is plain to see. The Yahoo! Web site gets an average of six million hits (visits) per day, from a base of approximately ten million users. It was, in 1996, the second most popular site on the Internet, after Netscape. The major competition for Yahoo! comes from other search engine firms, such as Lycos and Excite, and so-called metasearch engine providers like Metacrawler, which submit searches to multiple sites. The service philosophy at Yahoo! is to minimize the amount of time that users must spend at its site. According to the engineering vice president at Yahoo!, "As a search product, we know the user really wants fast access to other [home] pages, not ours. Hence we keep the graphics content of our pages low so that they will load faster."[15]

Each of the thirteen software engineers working in development at Yahoo! tends to be allocated full time to a specific project, with most projects carried out by a small team of people. Whenever a project nears completion, however, additional engineering resources are often pulled across from other areas to help get the product out the door. "We manage priorities, not projects," said the vice president of engineering. "Over the short term, we can assume the number of people available to a project is fourteen, including me."[16] The development team must therefore be flexible across as well as within projects. Team members are not wedded to the particular projects for which they have been given primary responsibility.

Such flexibility is not just related to allocating tasks. Given the nature of the product and the small size of the team (which can be a single person), Yahoo! engineers must possess a mix of skills. They must be broad enough to recognize the market requirements of concepts they are developing and deep enough to execute these requirements in robust software code. Engineers are also encouraged to pursue spontaneously new opportunities for additional services. A few weeks before the My Yahoo! project was due to go on-line, for example, the chief developer, needing a break, spent the hours between midnight and 6 A.M. developing a new sports page to provide soccer news and scores from the European championships just getting underway in England. By 9 A.M., after a short demonstration to the first VP to arrive, the service was put on-line on the Yahoo! home page—and became the most popular page over the four weeks the competition ran, with more than one hundred thousand hits per day. Indeed, when the Reuters link to the live scores went down one day, one of the Yahoo! founders (the only developer around with some spare time) jumped on a terminal and began typing in results from the wire link.

The Yahoo! development process is similar to Netscape's in that it emphasizes a slow release of software to users as the product becomes progressively more robust. Early versions of new services are first put on-line for internal use only. Given the development team's technical skills, these trials expose any major technical flaws in the product and provide additional suggestions for ways in which functionality could be improved. Once these changes are made, the service is put on-line, but without links to highly frequented parts of the Yahoo! Web site and without any promotion. At this stage, only the more technically aggressive users are likely to find and use the service. This exposes the service to rigorous external testing without revealing it to unsophisticated users who might get frustrated by a slow, incomplete, or error-ridden early version. Finally, at the official launch, the product is heavily promoted, normally by providing a direct link to the site from the Yahoo! home page. The product is thus finally exposed to volume usage.

Scalability is a major technical challenge for development projects for which potential usage is uncertain. At present, Yahoo! meets its processing needs with just a small number of small and inexpensive servers. Such low investment requirements for each machine mean that capacity can be smoothly scaled to meet demand rather than being added in large chunks. This gives benefits in both flexibility and the ability to experiment. The vice president of engineering described the approach:

> We have to be really concerned about volumes. If we promote a service on our home page, it will be seen by five million people each day. Even if only a small number "click the button," we will get hundreds of thousands of hits. In a similar manner, too much early success can be bad. Overloads slow the system before you have a chance to build capacity. Then you lose the early adopters, who might migrate to a faster service.
>
> Adding capacity in small chunks, however, gives us added flexibility. Our Web site setup works just like a spigot valve. If we want to test out a product on several thousand users, we put it up on the home page on only a few machines. As we reach the required volumes, we can also turn off the promotion for a given machine.[17]

Apart from helping to quickly scale services in response to demand fluctuations, the Yahoo! setup also has benefits in more effective experimentation. These stem from the ability to run multiple versions of the same software across different PCs in its network, varying the features on each. Doing so allows Yahoo! engineers to put slightly different versions of products on each machine and then track the results. They can, for example, load different feature sets onto different machines and track which attracts more repeat customers. With this setup, Yahoo! can also identify the customers that try a new service. By promoting a new product from a specific page in the main directory, users are filtered before they gain access to a new product.

The development process at Yahoo! is illustrated by the My Yahoo! development project.

The My Yahoo! Project

The My Yahoo! service is an individualized home page that presents users with a set of features, stories, and news items based on the

FIGURE 9-4

Development of My Yahoo!

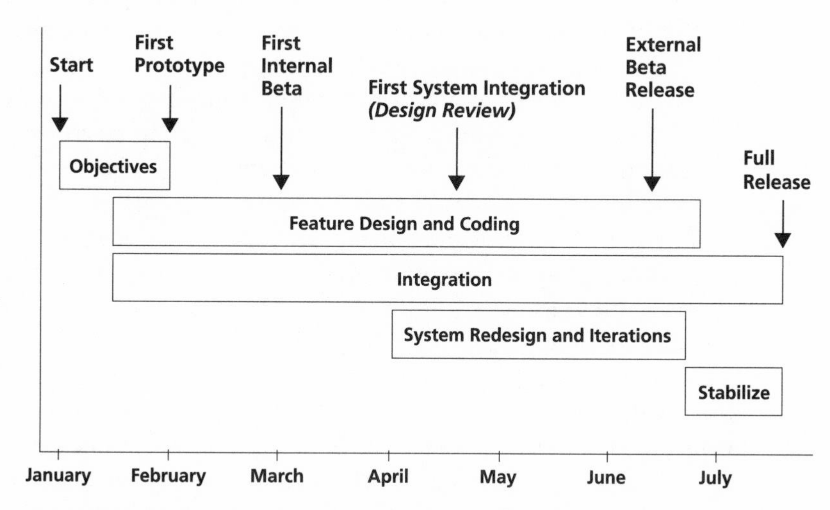

SOURCE: Alan MacCormack and Marco Iansiti, "Living on Internet Time: Product Development at Netscape, Yahoo!, NetDynamics, and Microsoft," Case 9-697-052 (Boston: Harvard Business School, 1996).

interests that a user has previously registered. The project was formally started in January 1996, with an initial specification consisting mainly of a list of features drawn up by the company founders.

Initially one person was allocated full time to the project. Although he was one of the newer recruits to the Yahoo! development team, he had five years of experience with LSI Inc., developing and using CAD software. By early February, he had produced a prototype of what the product would look like, comprising ten to twelve example HTML pages. With this approved by the founders, he began coding, and by the end of March the first (partially) working version was complete and made available to Yahoo! employees.

During March and April, several competitors (for example, Microsoft, Netscape, and Pointcast) brought out products similar to the one Yahoo! had planned. Yahoo! developers took time to evaluate these offerings, comparing them to their own. The marketing manager noted their conclusions:

> We thought they were rather dull and decided that they weren't really adding significant value over a basic search service. The problem was

> that they encompassed all the features that we had been planning to develop for My Yahoo! We needed a major rethink. . . .[18]

Subsequently, Yahoo! planners conducted a review of possible additional features and configurations and adopted a more aggressive blueprint to make the product more exciting.

At around this time, the VP of engineering decided to rewrite the original design to make it more robust to scaling up for higher volumes (this involved using a different language). Hence, during April and May the team was expanded to two full-time people, both developing new features and porting the code to a more robust platform. With a target date of June 17 for a soft launch (that is, a release to the outside world but with no promotion) the team was increased to three people at the beginning of June, and the first version of the new system, complete with new code and features, was integrated. This scaling up of the team late in the project produced major changes to the software, as the feature set was refined. A significant portion of the code was changed during the last four weeks of the project. The team developed a completely new stock-quote system and search engine, for example, to better match the service's requirements.

My Yahoo! was officially launched and promoted on July 15, 1996.

NetDynamics

NetDynamics provides development tools to help design Web pages with links to a company's internal databases. The product is differentiated by being extremely easy to use and by being extremely open (it supports Internet and database software products from any vendor, for example, and both Microsoft's Explorer and Netscape's Navigator browsers). Indeed, it is so easy to use that it takes only minutes to design a simple Web page that links to an internal database.

The company's first product, Spider 1.5, developed for the UNIX operating system, was launched in August 1995. It soon became clear, however, that the rapid growth of Microsoft's Windows NT operating system would require NetDynamics to provide a product that also ran on the NT platform. Late in 1995, it was decided that rather than trying to port across the original design to the new operating system, it would be better to develop a completely new version for both systems. Senior managers felt that the original design was neither robust nor comprehensive enough to sustain the product's early success.

NetDynamics 2.0

Design work for the new product, NetDynamics 2.0, started on December 1, 1995. To staff the project, seven engineers were recruited, each with a background in developing software for Windows-based systems. These seven joined the two engineers who had developed the original product to form the initial development team of nine. The initial target set by the CEO was to have a product ready by the end of February 1996.

During December, the team developed a prototype of the product (a mock-up of potential pages using Visual Basic). Existing customers were brought in and shown the prototype before being given a list of potential options and features and asked to rank their importance; this was done at an extremely detailed level to provide input to the design specification; for example, they were asked to name the specific database connections that were most important to them. The results from this exercise were tabulated on a spreadsheet and used to inform design tradeoffs. As a direct result of this exercise, the new product was designed to emphasize security aspects, a major concern of users because of the access that the tool provides to a company's internal databases.

During this time the most important design objectives were discussed and the system blocked out. The schedule did not allow enough time to develop a full HTML editor, for example, so it was decided to make the program general enough so that users could employ any of the commercially available editors. Besides making tradeoffs concerning the functionality of the new product, the team also had to decide which aspects of other vendors' products they could support adequately (for example, Oracle's command codes).

The single most important technical decision made during this period was the choice of scripting language. By the end of December, the company had reviewed five alternatives: a proprietary language (nicknamed "E"), Visual Basic (thought too elementary), C++ (a popular language, but with drawbacks due to the compilation time required, which would lengthen customers' development cycles), Python, and Java. The choice reduced to either "E" or Java, mainly because these would result in the maximum speed and simplicity with which a customer could develop an application. Making such tradeoffs was aided by referring to the vision characterizing the original product—to emphasize ease of use and to adopt emerging standards on an open platform.

While Java was ultimately selected, there was significant controversy about this choice, especially given Java's relative immaturity at the time of the decision. NetDynamics' chief engineer described the dilemma:

> We knew Java was going to be big, but it was still only available as a 1.0 beta version. This meant the development tools which went along with it were either terribly buggy or nonexistent. Subsequently, we had to develop many of our own development tools.[19]

To make the decision, engineers at NetDynamics spent weeks experimenting with various language options, trying to become as comfortable as possible with the potential and the risks of each. Since they were highly experienced in the other approaches, the engineers spent the bulk of their time familiarizing themselves with Java. They began by writing simple routines and slowly graduated to complex programs. They experimented with available programming tools and debated the pros and cons of each language intensely. Making this decision took about a month.

Working with other companies' beta software was a continual problem during the development cycle. Much of the work carried out with Java unearthed bugs that even its originator, Sun Microsystems, had not known about. In order to demonstrate that the bug was in Java, not their own code, the NetDynamics engineers often had to develop small test programs to expose the bug. On one famous occasion, with the team using a fast beta version of Netscape's Navigator software, all the engineering systems at NetDynamics suddenly crashed. After several hours of investigation, the chief engineers traced the fault to a time bomb installed in Navigator to stop unauthorized use past a certain date.

In January 1996, with the major design decisions made, the team had to commit to a schedule. They developed a project plan that analyzed how long it would take to bring together the two components of the new product, Studio, the development environment, and Runtime, the application server (the program handling data flowing between the server and the database). The group was divided into two teams, with five members working on Studio and four on both Runtime and some networking software.

By January 15, the Studio team had started coding. The Runtime team required a little longer to work out design issues and began coding around the beginning of February. The decision to start coding, however, was controversial. The chief engineer wanted another month to make the design more robust. There were many heated discussions, most surrounding the list of fifty to seventy open issues that were still unresolved. Although most of these were minor details that could be tackled later, about ten involved major design tradeoffs that had not yet been addressed. Despite objections, however, NetDynamics' CEO pushed for an early start. As he put it, "Time is more important than being perfect."[20]

By the end of February, the Studio team had the program framework up and running (without Java or networking capability) and started usability testing along with the development of a suite of regression tests (which simulate the impact of users). Around March 10, the two component parts of the system were integrated for the first time. The pressure for this integration came from the CEO, who had scheduled a meeting with analysts in New York to demonstrate the product. The team programmed the demonstration version to work only with a Netscape browser and a Microsoft database. While this initial version was a little shaky (and some features, such as networking, were not yet complete), it did show off the capabilities of the product from end to end.

Once this working model was achieved, the software was integrated daily, beginning in early March. Each morning, a daily build was performed, and everyone who needed the most recent version of the program received a copy. The team also adopted configuration management software to ensure code was not overwritten when checking in new features. When bugs were found in the daily build, they were reported against the current version; a centralized system tracked and allocated these bugs to the design team. By the end of the project, the system had logged more than six hundred forty bugs found and fixed since the daily builds began.

On April 8, NetDynamics announced that the product would ship within thirty days. Even though the product was not ready, marketing also pushed for a training program in early April, which, due to the lack of documentation, had to be run by the engineers who developed the product. At the same time, the team was working with twenty beta customers to bring the product up to the required level of reliability. Some features, however, were still in a state of flux, as were some major design issues. It had been expected that this would happen, as the market was evolving rapidly. As the chief engineer concluded, "We have to expect things to change, and so do our engineers. They have to be prepared to retro-fit their work as we determine the feature set has to change."[21]

FIGURE 9-5

Development of NetDynamics 2.0

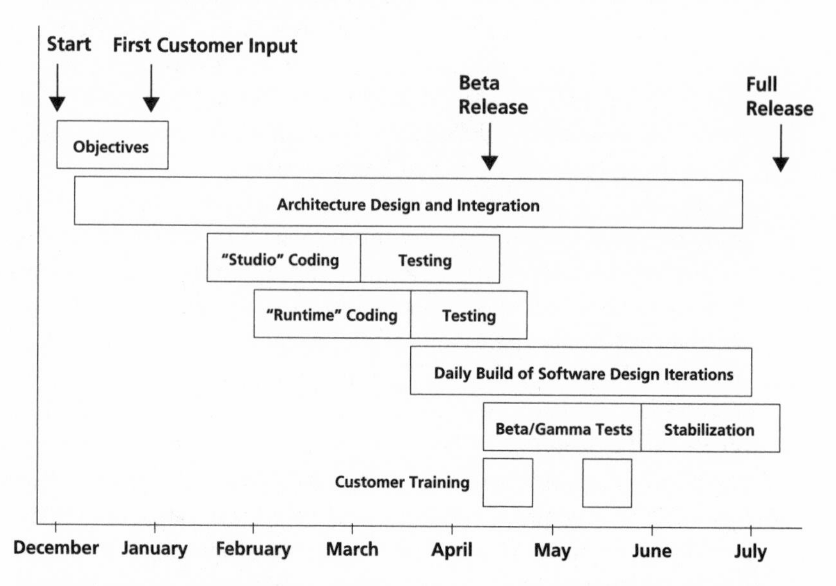

SOURCE: Alan MacCormack and Marco Iansiti, "Living on Internet Time: Product Development at Netscape, Yahoo!, NetDynamics, and Microsoft," Case 9-697-052 (Boston: Harvard Business School, 1996).

During late April it became clear that the product would be neither fully stabilized nor feature complete by the end of April. At this point, the engineering team decided to delay release. During May, they worked on fixing bugs, carrying out training with major customers, and performing another round of acceptance tests with gamma customers. During this period, NetDynamics learned to embrace those users that stretched their product to the limit, as this was often how they learned where those limits were. Customers often detected design flaws that NetDynamics would never have found until much later in the cycle. Providing rapid fixes, however, was critical to this approach. Customers accept bugs in software products only if the technical support to eliminate them can respond rapidly.

As the NetDynamics project approached release, the team began to move to focus on new development projects. The chief engineer described the challenge for the future:

> Development is a never-ending cycle of integrating new features and standards. Given the rate of change in the industry, we cannot rest.

> HTML is dying now, and Java applets are taking over. As soon as this product is released, we have to start developing again. It's worse because we are so "open." We have to support everything.[22]

"The Recovery": A Microsoft Update

Microsoft was late to enter the Internet environment; its senior management team was slow in recognizing the opportunities offered by the World Wide Web. It was not until the end of 1995 that Microsoft really began to focus on developing critical Internet products, such as a browser that would rival Netscape's Navigator. When Microsoft entered the browser market with its first release of Explorer, Netscape already had a dominant position and was way ahead in features and design.

Thus, at the end of 1995, Microsoft appeared to have missed an opportunity larger than the personal computer—yet it recovered from the strategic mistake with incredible speed. In six months, from the end of 1995 to the middle of 1996, it went from no presence in the critical browser market to offering a product that several industry experts claimed was comparable to or better than Netscape's Navigator. By August 1996, Microsoft still had a small share of the browser market, but its Internet business was gaining momentum. Netscape stock had lost 20 percent of its peak value. An architect in Microsoft's advanced technology group characterized Microsoft's response to the Internet challenge this way:

> Think of how far behind we were a year ago. It's true that we missed an opportunity to put our name out in front. But within a year we put out something that was better, and integrated from beginning to end. Imagine what someone else would have done.
>
> This is what we are good at at Microsoft. We will argue with each other and procrastinate until someone on the horizon paints a target on their chest, like Netscape. Then we go for the kill.[23]

Although it started behind the industry leaders, Microsoft already had the right product development philosophy in place. When Gates and the rest of Microsoft's senior management team finally acknowledged the need for a strategic shift, Microsoft's development groups were ready to take action. Its applications development process already shared many similarities with those at work at Netscape and Yahoo! It was founded on the rapid iteration of prototypes, early beta releases, and a flexible approach to product architecture and specification. The Internet version adopted was simply more extreme and even more intense—driven by Internet time. It was also more externally focused, making use of the instantaneous distribution and customer feedback potential offered by the Internet itself.

The process behind Microsoft's Explorer is founded on rapid iteration around betas, like that of Netscape's Navigator. It is more internally focused, however, and driven more by internal prototype releases than by external customer feedback. Microsoft will generally go through about half the number of betas per cycle that Netscape does, with two or three, rather than six or seven, major releases. The approach relies heavily on extensive testing through Microsoft's own Intranet, as is explained by one program manager:

> We try and be a bit more careful about staging betas than Netscape—we need a clear reference point for feedback; things get confusing if you release too many different beta versions in a short period of time. However, we build everything as frequently as possible. Doing a system build of Explorer only takes fifteen minutes. Then we release it for internal testing. Everyone around Microsoft is encouraged to play with it. You have to realize we are a big company now. Internal testing means that we release it to thousands of people that really hammer away at it. We use the product much more heavily than the average Web user. By the time we ship a version of Explorer it will have had much more test time than Navigator. This is because we have a huge Intranet, and we have been very aggressive about using it for this purpose.[24]

Microsoft's rapid recovery was enabled by a large reserve of skill and experience. The company has an incredible depth of talent, a vast reserve of individuals good at problem solving under high pressure. Once it decided to go after the Internet, the organization dedicated its best resources to the effort. One of Microsoft's early Internet champions describes a job listing for the Internet Explorer group.

> The job requirements were incredible. I still remember the internal job listing for a program manager spot on Explorer. You had to be a level 12, which is one level below business unit manager. You had to have at least a 4.0 average for reviews, which is almost impossible. You had to have shipped at least four products, and had to have been at least five years at Microsoft. The amazing thing is that these positions were

> staffed almost immediately! This is one of our greatest advantages. We have that kind of people floating around here. We are out to kill, and we have people here that can do this.[25]

Microsoft rapidly adapted to the turbulence of the Internet. It released three major versions of Explorer in nine months, and by the middle of 1996 had begun to threaten Netscape's dominance.

Patterns: Product Innovation on Internet Time

We have just analyzed four very different firms. The first, Microsoft, the leader in PC software development, is responsible for a wide variety of products ranging from word processors to encyclopedias, from video games to operating systems. The next three are start-ups, scarcely two years old: Netscape, developing Internet products aimed at broad, consumer applications; Yahoo!, developing consumer services; and NetDynamics, developing tools for relatively sophisticated firms that want to link Web sites to internal databases. Despite the differences, the development processes of all four companies resemble the flexible model described earlier, with emphasis on early integration of both external user requirements and internal product features. Microsoft was one of the first to discover and formalize the value of early integration and rapid iteration. The others adopted a similarly flexible philosophy but pushed it toward even greater levels of iteration, responsiveness, and customer integration. Microsoft, after conquering some strategic uncertainties, was then ready to learn from them and apply these lessons to improve its own internal processes.

The similarities among all cited development processes are striking. The concept development and implementation phases are highly overlapped; team members quickly translate product features into a functioning prototype. User input (internal and external to the company) is continuously integrated into the development process. The development proceeds via rapid iterations in which major changes are *expected.* Firms strive for an early integration of the system components. Once achieved, they build the system on a daily basis to evolve the product.

Generating Knowledge of Technology, Market, and Customer

The reactive nature of the projects described above is useful only because there is good information to react to. The value of flexible development therefore hinges on the quality of the process for generating knowledge about the interaction among technology, user needs, and market requirements. Unlike traditional development projects, where research on user needs provides occasional bursts of input to the development process, projects here use essentially continual feedback on critical features. Such feedback interacts strongly with the development process, greatly influencing daily decisions, technology choices, and tradeoffs. This is made possible by the nature of the Internet infrastructure. Companies can contact users, send a program, and receive feedback in a matter of hours.

Nowhere is this used more than at Netscape, where the release of up to six beta versions of a new Navigator product means that users help significantly to evolve the product, providing feedback on features that have been integrated into the existing version, and suggesting new features that could enhance the product. Up to 50 percent of the new code, features, and technology components integrated into a new release of Navigator are developed *after* the first beta version has been released. Netscape therefore has considerable flexibility to respond to changes in market demands within its product development cycle.

At NetDynamics, once the initial mock-ups were completed, lead users were brought in to evaluate the NetDynamics' functionality. At the same time, design tradeoffs were discussed and the architecture of the product outlined. Soon after the first integration of the complete system, the development team was working closely with twenty hand-picked beta customers who would stretch the limits of product performance. Having to run a training program during the critical April before their scheduled May release date also gave the development team "hands-on" experience of user problems and requirements. During May, additional gamma customer testing was used to refine the software.

At Yahoo!, external users were integrated into the development process at a later stage than at either Netscape or NetDynamics, since, being a service provider, Yahoo! believed that before a service is released to the outside world, it should be robust. Users who try a service once and have a poor experience with it are unlikely to return. Furthermore, once a new service is released, it is assumed that innovative features will be copied by competing firms. These factors tend to suggest delaying external testing to later in a development cycle, compared with software product providers.

To mitigate this potential disadvantage, Yahoo! relies on its internal staff to adopt the position of lead users during the development cycle. Furthermore, once a service moves to external testing, it is given a significant amount of exposure, ramped up over a very short period of time. To achieve this, Yahoo! relies on some thirty thousand users who have volunteered to become beta test sites. In developing My Yahoo!, they selected five hundred who were given passwords to access the new service. After several weeks, a soft release occurred, in which the product was released on the main servers without any promotion. The full release happened several weeks later.

Similarly, Microsoft obtains rapid, high-quality feedback within the development cycle through the firm's employees, who often play the role of lead users. Development engineers are expected to be extensive users of the technologies they are developing. They are also expected to be up to date with the functions and features of all of their competitors. The extensive internal feedback is then combined with more carefully staged external beta releases. This is done to organize market feedback and to limit the risk that imperfections in early product releases would damage the company's reputation.

Integration: Overlapping Conceptualization and Implementation

To be effective, of course, customer and market knowledge must be rapidly translated into decisions about technology and product features. Projects must respond rapidly to changes within the development cycle. There is little point in waiting until all the major design issues are resolved before beginning execution—they will never be resolved completely. Projects should therefore expect a significant amount of design work to be thrown away during the evolution of a design.

The My Yahoo! project illustrates these dynamics well. After three months, a partially complete version of the service had been developed. However, several competing versions of similar services had also been launched. In addition, the phenomenal growth in Internet popularity, particularly the Yahoo! main search engine, had cast doubts on whether existing software could support such usage. Thus, Yahoo! conducted a comprehensive review of the product and decided that major revisions were necessary. The software was to be rewritten so that, as volume increased, it would be more robust to scaling up. A new set of features and functions was also developed to better differentiate Yahoo's product. As the effort continued, and new software was integrated, the team found that parts of the Yahoo! main service also had to be redesigned to work with it. As noted, more than half of the software changed during the last four weeks—meaning that by launch date, the team had rebuilt the new service almost three times during the development cycle.

Product development at NetDynamics also illustrates this process. While many critical decisions were being made in early phases of the project, much uncertainty existed over the future evolution of the Internet. The critical first choice of using the scripting language Java was hotly contested, especially given limited information about its capabilities. The decision was informed, however, by the open philosophy ingrained at NetDynamics. Developing an emerging standard would take precedence over developing proprietary technology. Yet when coding started, there were still some fifty issues to be resolved, of which ten were major design tradeoffs. The decision to begin execution was taken not because these issues were unimportant, but because it was expected that the next few months would bring additional information with which to make the necessary choices.

Experimentation and Experience

This book has stressed that good decision making is built on knowledge—which is generated through experimentation and accumulated largely through experience. In the Internet world, the rapid obsolescence of knowledge makes experimentation the centerpiece of all development efforts. Reflecting this idea, companies in this field emphasize the building of working system models or prototypes early in a given project, as the three projects studied here demonstrate. This defines the user context and desired features of the product. It is a critical milestone, as it lays foundations for a software architecture, sets priorities for feature development, creates a vehicle for user feedback, and allows some of the most important tradeoffs to be addressed. While the prototype does not have to be fully functional, it must reflect the essence of the product in terms of how users will interact with it.

Although early prototyping is common to all firms, the sophistication of early prototypes varies considerably. At Yahoo!, a demonstration version of My Yahoo! was produced one month into the six-month development cycle and consisted only of a mock-up of the HTML pages that would form the user environment for the service. At NetDynamics, a mock-up of the user development environment was also produced and shown to customers during the first month of the project, to inform decisions about desired features. At Netscape, however, the first integrated prototype of version 3.0, the internal Beta 0 release, was relatively comprehensive. Many new functions had been integrated, representing over 50 percent of the new code required. Even with this sophistication, the beta 0 version was still released early in the new development cycle, six weeks after the architecture discussions had started.

As the pulse of the project, prototypes punctuate the effort by integrating architecture and technology into a coherent package. They are essential to discovering interactions between various parts of the program. Once the code has been debugged, the prototype provides a baseline for the addition of new features or refinements. These are checked in by the engineers, who then run a number of local experiments to ensure that the new code does not interfere with the functioning of the other parts of the prototype. The cycle is repeated at least daily.

Experimentation is not everything, however. When I asked project members from each company about the ingredients most critical to a successful project, they consistently pointed to the experience of software engineers and managers. Their experience was deemed crucial to guiding the experiments, to navigating through the complexity of the technology and user environment. The engineering and operations vice president at Yahoo! underscored this point: "Experience is essential. It is the only thing that lets you see how the whole system works. It is critical to make the right tradeoffs as the project evolves."[26]

Ultimately, Microsoft's speedy recovery is linked to its focus on experimentation combined with its incredible depth in experience and skill. Once the company's strategic shift away from its proprietary on-line service was finally made explicit, this talent and its associated R&D philosophy was turned loose on the Internet, with impressive results.

Mastering Reaction: Integrating Technology and Market Streams

The vast uncertainty of the Internet environment has deeply influenced the nature of research and development. Gone are the days of clear objectives, frozen specifications, and proven technologies. If we wait until all uncertainties are resolved, the market opportunity will disappear. Leading firms have acknowledged the need for building flexibility into product development and have developed processes characterized by rapid and extensive experimentation and iteration in objectives and specification—all founded on a solid base of experience and skill.

The responsiveness required by the emergent Internet environment has sharpened the need for integrating technology. Integration, in this environment, means creating flexible links among a vast variety of technological possibilities and the rapidly evolving context of application. This goes well beyond the now-established emphasis on cross-functional integration in development project teams. Let's imagine being at Netscape halfway through the development of next generation Navigator and realizing that our key competitor has just released a beta version of their browser that, unlike all previous versions, is highly customizable. To respond effectively, we need to know where to find the necessary technology, say, software algorithms for creating and placing new menu buttons. We also need to know where to find the necessary programming skill, the necessary usability experts, and the necessary designers who will translate the technology into a distinctive and usable application. We need to know how to test the new interface with internal and external users, and we need to know how to interpret these results. We need to know how customizable interfaces influence the vast variety of Internet standards. Then we have to make a critical decision—Do we respond by changing the current product, which is expected to ship next month, or do we respond in the next product, which we expect won't be out for at least another six months? Making this decision is so painful that we might even wonder whether our competitors had introduced their beta because they wanted to throw us off our stride or because they really intend to go down the customizable path.

Such needs require going beyond traditional notions of product development activities as being focused and independent projects. They require integration between highly diverse knowledge domains, scattered across projects, organizations, and customer bases. Developing products for the Internet is more like managing the interaction between two streams, characterizing the evolution of technological possibilities and market preferences (see Figure 9-6). The challenge of product development is deciding when a given set of technical options should be fished from the first stream, integrated into a product, and offered to users in the second stream. The bundling can be done either as a beta prototype or as a final product release. In either case, the product is not a final outcome but simply the next step in the interaction between the two rapidly evolving environments. The feedback generated from the market's response to the product will greatly influence its next release.

FIGURE 9-6

Integrating Technology and Market Streams

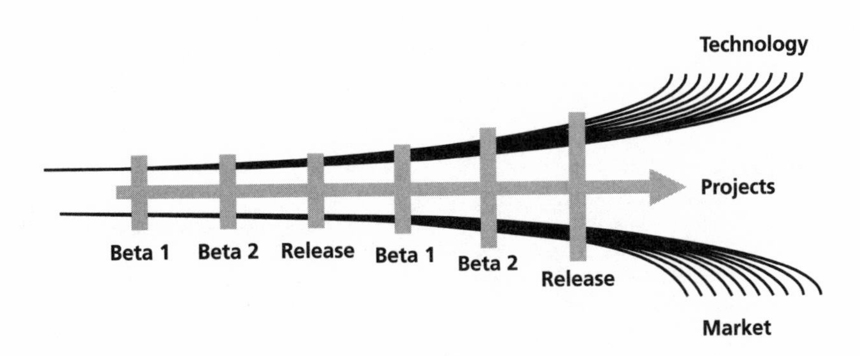

This model implies a shift toward externally available skills and knowledge. It is impossible to develop the needed breadth of options internally. Even Microsoft, with its extensive internal expertise, went to an outside firm to license a browser program, which accelerated the introduction of the first version of Internet Explorer. The external focus necessary for this environment is thus not restricted to input from customers, but includes access to a variety of sources of technical and market know-how, ranging from academic institutions like MIT's Media Lab to the vast variety of software startups in Silicon Valley. The emergent model thus implies deep internal technology integration capabilities coupled with access to external sources of knowledge, such as relevant research or engineering skills. In other words, both technology and market streams are largely external to the firm, and the core capability of the firm is the interaction between the two, through technology integration.

The need for integration therefore goes well beyond achieving good communication within a well-defined project team. It implies having access to the technology and market streams, which are in large part external to the firm. This means having knowledge about a diverse variety of external technologies and access to diverse customers. It implies knowing where to find needed skill sets and resources, internal or external to the firm. It implies having access to tools and methods for broad experimentation and having a deep enough understanding of the application context to interpret their results. This requires integration way beyond the local project boundaries, reaching into external technology bases and customer needs, accessing talents and knowledge disciplines that were not even expected to be relevant when the first product specification was written.

Notes

1. This section is drawn in part from two Harvard Business School cases: Geoffrey K. Gill and Marco Iansiti, "Microsoft: Office Business Unit," Case 9-691-033 (Boston: Harvard Business School, 1990), and Ellen Stein and Marco Iansiti, "Microsoft: Multimedia Publications (A)," Case 9-695-005 (Boston: Harvard Business School, 1994). Interested readers might want to refer to several other references, including Cusumano and Selby (1995).
2. The "developers" at Microsoft were the software engineers responsible for writing the software code.
3. Geoffrey K. Gill and Marco Iansiti, "Microsoft: Office Business Unit," Case 9-691-033 (Boston: Harvard Business School, 1990).
4. Ibid.
5. Ibid.
6. Ibid.
7. Ibid.
8. Ibid.
9. Ibid.
10. Ibid.
11. Alan MacCormack and Marco Iansiti, "Living on Internet Time: Product Development at Netscape, Yahoo!, NetDynamics and Microsoft," Case 9-697-052 (Boston: Harvard Business School, 1996).
12. Ibid.
13. Ibid.
14. See Clark and Fujimoto (1991), for details on the argument that the faster the speed the higher the productivity of resources.
15. MacCormack and Iansiti, "Living on Internet Time."
16. Ibid.
17. Ibid.
18. Ibid.
19. Ibid.
20. Ibid.
21. Ibid.
22. Ibid.
23. Ibid.
24. Ibid.
25. Ibid.
26. Interview conducted by author during field research from 1990 through 1995.

Rethinking R&D

chapter 10

in a Chaotic World

THE TWENTIETH CENTURY has witnessed the triumph of the science-based enterprise. The imperative to understand and harness science motivated the establishment of industrial laboratories to conduct research on fundamental scientific phenomena. This was critical to the explosive post–Second World War growth of such firms as IBM, AT&T, Philips, Dupont, and General Electric. These, and many other firms in many other industries, drew much of their strength from the excellence of their industrial research, which allowed them to dominate their markets worldwide. By the 1990s, however, the very core of the science-based enterprise had been shaken (see, for example, Rosenbloom and Spencer, 1996). Every major industrial R&D laboratory, including AT&T Bell Laboratories and IBM's T.J. Watson Research Center, was restructured, downsized, or simply shut down. There was enormous pressure to justify research budgets by proving their direct impact on the bottom line.

This upheaval was not due to any fundamental maturity in the technological base; if anything, technical challenges like those involved in further reducing semiconductor dimensions appeared more difficult than ever before. Simple improvement paths had been exhausted, and further evolution depended on mastering novel approaches in domains ranging from optics to plasma physics. Even in software, Internet search engines or three-dimensional graphics display programs drew upon the latest recent research in linguistics and mathematics to increase speed and efficiency. Indeed, during the same time that industrial R&D labs were under siege, investment in research, new products, and capital equipment across the computer industry actually *increased* to unprecedented levels. Technology had not become irrelevant; it had become even more central to competition. And new technology was based on scientific developments, ranging from shallow trench isolation at IBM Microelectronics to distributed shared memory architectures at SGI. Both advances had a strategic impact on the competitiveness of these firms.

Evolving Challenges for the Science-Based Firm

But if technology was central to competition, why were R&D institutions undergoing such struggles? The research outlined in this book provides insights that help answer that question. These suggest that while the impact of science and technology on business was perhaps greater than ever, the nature of the challenges facing the science-based enterprise had deeply changed.

First, knowledge of scientific foundations had become increasingly pervasive. By the mid 1990s, graduates from leading universities populated the R&D organizations of companies all over the world; their expertise in science and technology and their familiarity with the latest research had laid the foundations for innovation in nearly all competitors, regardless of size. This meant that mastery of science was necessary, but no longer sufficient to guarantee competitive advantage. In addition, this diffusion of science had enabled the growth of a widening range of suppliers likewise familiar with the latest innovations in software, computer architecture, semiconductor process equipment, or materials science—and likewise located around the globe. As such, the mastery of all the science base necessary no longer had to be internal to a single firm.

The science base relevant to a single product had also grown exceedingly broad and complex. Leveraging external scientific knowledge had become critical, since no single organization could develop all its options internally any longer. A workstation, in its entirety, made use of almost every field in the physical sciences and mathematics

—from the physics of nuclear decay (relevant to DRAM design) to the mathematics of graph theory (relevant to its software routines). Given the enormous breadth of this range of fields, it had become impossible for a single firm to research every relevant discipline, as IBM was doing during the 1970s and 1980s.

Moreover, the environment surrounding these research institutions changed increasingly quickly and unpredictably. By the mid 1990s, the computer industry was highly, if not wildly, uncertain—as manifest in such market drivers as the penetration of the Internet into homes or the price of DRAMs. Uncertainty in the marketplace was coupled with uncertainty in the technical base; this ranged from the potential of ultraviolet lithography to the robustness of Java as an Internet scripting language. All this uncertainty deeply interacted with the relevance of individual scientific disciplines and knowledge bases. This uncertainty repeatedly challenged any technical choice and pushed organizations toward faster and increasingly flexible development projects.

The Need for Technology Integration

These trends changed the demand on R&D in the science-based enterprise. In the traditional R&D organization the goal was to create knowledge of a few selected disciplines (domains), which was then transferred to a stable process for creating a clearly defined product. Today, what is traditionally thought of as R&D must provide relevant knowledge to an incredibly complex and uncertain context. The source of this knowledge is more diverse than ever, and it is largely external to the firm. This book has argued that in such an environment what confers advantage is the capability of choosing among knowledge domains and integrating them, so as to conceptualize products (and services) that fit into a rapidly changing environment.

This, I would argue, is fundamental to explaining the struggles of industrial research of the 1980s and 1990s. Industrial labs grew out of a need to separate a research organization from the incremental requirements of day-to-day operations, so as to create or discover novel technological concepts. Yet these labs have more recently experienced enormous pressure to swiftly translate a concept into reality. These aims are often completely antithetical. The first implies a broad outlook, as scientists pursue the potential of a variety of possibilities; the second demands a single focus, implying choices and integration. Because the traditional lab had not been structured to accommodate the latter aim, a critical gap was exposed, leading to delays, inefficiencies, and market failures. My research suggests, however, that a technology integration capability can fill this gap, substantially improving the impact of research and the associated competitiveness of the firm.

My research shows that while R&D organizations were being overhauled, effective enterprises were hardly pulling the plug on science. The evidence repeatedly links revolutionary projects to a strong research base. What effective firms had recognized was that the application of science had become vastly more difficult and that they needed a process that selected from an expanding variety of possibilities. Effective firms were not doing away with research; they were getting better at integrating its output. The value of research had, if anything, increased. This was more and more an option value, however, and the decision of which option to choose had grown in subtlety and impact.

Improving the Process

These improvements in the capability to integrate new technologies did not happen by accident. They happened because managers recognized the technology integration process as a critical lever in managing technological change. And organizations thus invested in the knowledge generation, retention, and application processes needed to make the process effective.

The evidence in this study implies that investment in knowledge generation through experimentation often drove improvements in technology integration capability. The most notable example was the significant improvements made by U.S. semiconductor firms over the last ten years, when the experimentation capacity per project grew in some cases by more than a factor of ten. Intel, IBM, and Texas Instruments built massive facilities to provide their projects with the necessary tools. Similarly, Silicon Graphics built one of the largest supercomputer centers in the world to support its project teams, and it nurtured deep relationships with software vendors to guarantee that the simulation software used would push the absolute state of the art.

Effective organizations also focused on the retention of knowledge through experience as a clear, long-term objective. In some cases this was relatively easy to do. In Japanese companies, lifetime employment practices mesh well with the need to retain experienced integrators. But other organizations stepped well outside accepted practice to retain the knowledge needed. Silicon Graphics and Intel both built a deepseated tradition for retaining their technical talent over time. In Silicon Valley, this is extremely difficult to accomplish, and during my research I saw repeatedly how senior management would become directly involved in career-path decisions for both managers and engineers. Compensation and incentives were also structured to meet the need of retaining critical employees.

Moreover, these organizations also saw that the knowledge gathered was applied in a systemic fashion—and at the right time in the evolution of a project. This required making sure that a focused group of individuals was working on technology integration decisions *before* the project was forced to commit to its basic technical approach. The group needed to have enough time and resources, as well as the requisite experience base and experimentation capabilities. Not surprisingly, this study repeatedly found that an increased focus on integration was coupled with significant organizational and procedural changes. At IBM, technology integration expertise was centered in their newly formed Semiconductor Research and Development Center organization. At Microsoft, improvements in integration were associated with the evolution of their program management function and with significant changes in the definitions of gates and milestones.

Strategic Impact

Thus, a group of firms in the semiconductor industry responded to fierce competition and significantly improved their ability to introduce new generations of process technologies. A different group of firms in the software industry rapidly built the capability to develop products in the Internet environment. These achievements built on, but moved beyond, the solid foundation of science created during the 1980s and 1990s. Improvements in technology integration capability enabled these organizations to leverage this foundation much more effectively. Doing so involved a substantial investment in infrastructure, indicated by a higher capacity for knowledge generation through experimentation, as well as by a significant change in organizational process, which provided more effective knowledge application and retention over time.

These companies made such changes because they recognized that an effective process for the integration of technology can have a *strategic* impact on the firm—first of all because technology integration can be instrumental to the execution of a challenging strategy through effective projects. Intel's dominance of the microprocessor environment is enabled by its development of speed and efficiency as well as by the resulting yield of its manufacturing processes. Inefficiencies and delays observed in some mainframe projects were repeatedly associated with drastic consequences, such as shifts in market focus or even industry exit. Even the traditional dominance of IBM was seriously threatened.

But differences in project performance are not the only critical outcome of an effective technology integration process. Experimentation performed in Netscape's projects was instrumental to defining the basic boundaries of its business. Its managers consistently used feedback from beta releases to conceptualize and justify fundamental shifts in project focus and strategic direction. It was through discussions about product specifications and the feedback from simulations that Silicon Graphics managers realized that commercial markets were a desirable focus. The technology integration process thus expanded the views and options of critical decision makers. The evidence in this book suggests that the investigation of multiple technological possibilities, and of their interaction with the external context of the firm, can lead to the recognition of strategic threats and opportunities.

Technology integration can also have a critical impact on the firm because it enables the quick validation of the feasibility of a possible strategic shift. Silicon Graphics' move toward distributed shared memory architecture looks impulsive and reckless—until one considers the impact of the experience of the Stanford research group and of the considerable simulation capability at the project's disposal. The extensive discussions at the senior management level, the careful nurturing of the relationship with Stanford, and the investment in sophisticated simulation tools turned a reckless strategy into a bold one. By the time the decision was made, a preponderance of evidence had accumulated validating the original claims. Similarly, it was easier for NetDynamics to bet on Java as the language of choice once the project's feasibility had been quickly investigated through rough prototypes.

Finally, technology integration has an impact on the organization because it provides flexibility in projects, enabling them to adjust to changes in context on the fly, as I saw with the software projects conducted at Netscape, NetDynamics, and Yahoo! And adaptability at the project level enables commitment at the strategic level. It is much easier for a general manager to establish a firm goal if he or she can count on a project-level process that is flexible enough to overcome emerging and unpredicted obstacles. This is why Silicon Graphics is known for canceling many fewer projects in midstream than do its competitors.

Ultimately, the mastery of technology integration is important because it creates the capability to manage technological change. It enables an organization to navigate between technological trajectories, fighting inertia and adapting to and influencing an uncertain market context. And reacting to uncertainties in technology and market streams has become the quintessential skill for life in the increasingly chaotic world of the science-based enterprise. This is also not an accident. Ed McCracken, SGI's CEO, expresses it this way:

> There is only a handful of firms left that are defining the direction of the workstation industry. . . . The source of our competitiveness in this industry is our ability to manage in a chaotic environment. But it's more proactive than that. We actually help create the chaos in the first place—that's what keeps a lot of potential competitors out.[1]

Notes

1. Ellen Stein and Marco Iansiti, "Silicon Graphics, Inc.," Case 9-695-061 (Boston: Harvard Business School, 1995).

Appendix I

Technical Analysis of Processor Modules

The design of an advanced computer processor interconnect substrate provides an ideal environment for studying technological yield. (Figure I-1 shows an example of a processor interconnect substrate.) Developing a new processor substrate is a challenging and complex effort, involving work in basic scientific research as well as a number of subtle product and process design considerations. Moreover, the functional performance of a processor interconnect system is well defined and is dependent on physical parameters via a simple scaling model.

A computer processor comprises two fundamentally different types of devices: the logic gates, packed in integrated circuits, and the interconnect system whose heart is the multilayer substrate shown in Figure I-1. While the integrated circuits perform the actual logic functions of the computer (for example, comparison, addition, and storage), the interconnect system links these gates to each other so that a complex set of instructions can be communicated and executed.

The drivers of processor performance (usually measured by the number of instructions per second) are shown in Figure I-2. A number of basic paths aim at improving processor performance. One can work on the hardware (logic and interconnect system, whose contribution is measured by the cycle time) or on the software (instruction set characteristics and number of cycles per instruction). The critical contribution of the processor interconnect system is indicated by the thicker line in Figure I-2.

Interconnect system performance is measured by its speed.[1] One great advantage of focusing on this type of device is that the performance vector $\underline{P}$ only has one critical dimension, P_I. Interconnect speed can be measured precisely as the average time that an electrical signal spends traveling between integrated circuits during a single instruction cycle. The shorter this delay time, the higher the

FIGURE I-1

Typical Mainframe or Supercomputer Module

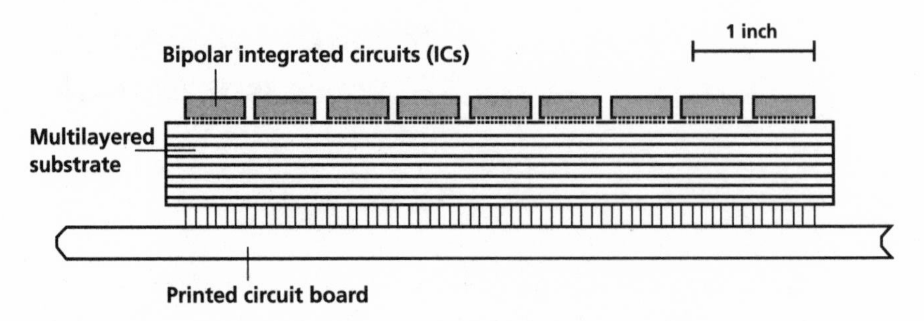

NOTE: The multilayered substrate is the focus of the technical analysis in Chapter 5.

SOURCE: Adapted from Marco Iansiti, "NEC," Case 9-693-095 (Boston: Harvard Business School, 1993).

performance of the interconnect. Interconnect performance, therefore, is defined as $P_I = 1/t_d$, where t_d is the delay time. The higher P_I, the higher interconnect system performance.

The delay time depends on a small number of physical parameters (see, for example, Dohya, Watari, and Nishimori, 1990; and Tummala and Rymaszewski, 1988). The first is the dielectric constant of the material, ε. The dielectric constant affects the speed of the signal between integrated circuits. Signals travel at the speed of light, and the speed of light on a given material is given by $c/\sqrt{\varepsilon}$, where c is a universal constant ($c = 3*10^{10}$ *cm/sec*). The dielectric constant can have very different values in different materials.

Apart from the speed of the signal, delay time will also depend on the average distance the signal must travel between logic gates. This is given by the square root of the density of the circuit, given by g, measured in the number of gates per square centimeter. Since the time elapsed in traveling between two points is given by the distance divided by the speed, the average interconnect delay time t_d in an ideal system will be given by the following equation:

$$(t_d)\text{ideal} = (1/c)\sqrt{(\varepsilon/g)} \quad (1)$$

Technological potential TP(ε,g), is given by $1/(t_d)ideal$. Therefore,

$$TP(\varepsilon,g) = c\sqrt{(g/\varepsilon)} \quad (2)$$

TP is an upper bound for performance, which is limited by the

FIGURE I-2

Computer Performance Drivers

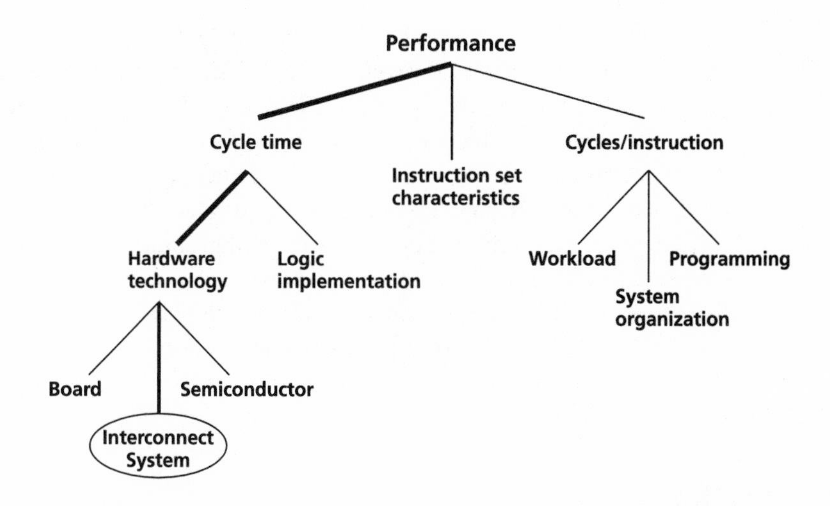

NOTE: The impact of the interconnect system on speed is highlighted. In a typical mainframe from the late 1980s, the interconnect system contributes about 50 percent of the total delay time in the processor and is therefore critical to performance.

SOURCE: Adapted from Marco Iansiti, "NEC," Case 9-693-095 (Boston: Harvard Business School, 1993).

speed of light in a medium. This simple analytical model is derived from basic principles of physics. Its elegance reflects the functional simplicity of an interconnect system. The only task is to carry electrical signals, and the faster a system can do this, the better its performance will be. The model is used throughout the industry for technology planning and product specification (see, for example, Dohya, Watari, and Nishimori, 1990; and Tummala and Rymaszewski, 1988). Figure I-3 illustrates the progression of the technological potential as a function of time for different technologies as described in a technical paper written by NEC's R&D organization.

The model highlights several possible fundamental knowledge domains for improving the performance of a processor interconnect system. First, an R&D organization can work on developing materials with lower dielectric constant. Second, it can work on improving the density of the system, for example, by developing superior technologies for patterning interconnect wires. These types of improvements capture the trends shown in Figure I-3. Going from single to double-sided printed wiring board technologies, for example, achieved a twofold improvement in gate density. Adding polyimide layers to ceramic packages achieved a substantial improvement in dielectric constant.

Many ways to improve actual product performance are not included in the model, however. Local defects in the interconnect pattern will influence the local capacitance (and inductance) of the wires, for example, which will cause signal delays. Minimizing such defects increases actual product performance (leaving the technological potential unchanged). While degrading product performance, these (and other) considerations cannot be modeled precisely *ex ante* in a clear, closed-form fashion. Their improvement, therefore, is based largely on experiential knowledge and experimentation, the latter through sophisticated simulation models or through the fabrication and test of physical prototypes. The impact

FIGURE I-3

Technological Potential Values in Processor Interconnect Systems

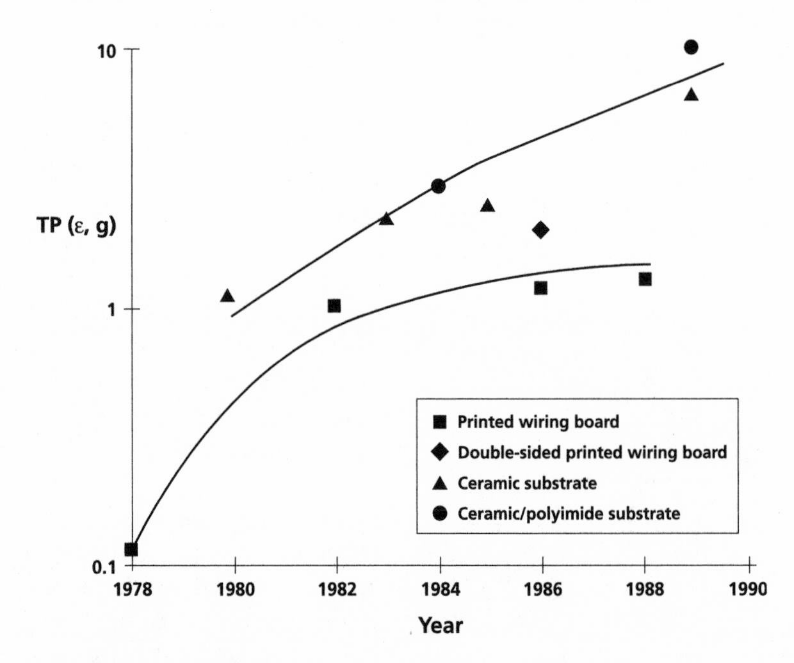

NOTE: The lines indicate the trajectories for printed wiring board and ceramic technologies.

of these other factors is included in the yield function, *TY*, given by the following equation:

$$\mathbf{TY = 1 / [t_d\ TP(\varepsilon, g)] = P_I / TP(\varepsilon, g)} \qquad (3)$$

In summary, the simple closed-form upper bound model *TP(ε,g)* described here links fundamental physical parameters to system performance in advanced processor interconnect systems. I have argued furthermore, that other considerations also impact product performance. These will be captured by the technological yield *TY*.

Notes

1. In the design of interconnect substrates in high-speed computing there is only one primary performance measure, and other performance variables are not as critical. In other words, the substrate is designed to maximize speed given an acceptable level of thermal dissipation, chip protection, and reliability. In the processor designs included here, speed is consistently the only binding constraint—levels of heat dissipation are enhanced in all products by the design of water-cooled metal pistons, so that the thermal conductivity of the substrate itself is not a critical performance measure; reliability in all cases is high enough that the system will never break down because of substrate problems.

Appendix II

Regression Analysis

Tables 1 and 2 in Chapter 7 are drawn from the regressions in the following tables. While the models selected are exactly the same as those shown in Chapter 4 regressions for performance, the results are quite different.

TABLE II-1

Determinants of Revolutionary Performance (Technological Potential)

Dependent Variable	Logarithm of Technological Potential
Intercept	–316*** (26)
Time	0.159*** (0.013)
DRAM	–0.788*** (0.114)
Project resources (PY)	–0.00097** (0.00026)
Experimental capacity (wafers/week)	0.00019** (0.00008)
Experimental iteration time (weeks)	0.0016 (0.019)
No previous project experience	0.23 (0.28)
Research experience	0.27*** (0.091)
F-test	93.53***
R-squared	0.969
N	29

The dependent variable is the logarithm of technological potential, which is defined as *(number of layers/linewidth)*^{2}. Time is a continuous variable corresponding to the time of process introduction. The DRAM dummy is *1* if the project was aimed at the DRAM segment, and *0* otherwise. A * means significance at 10 percent level, ** significance at 5 percent level, and *** significance at 1 percent level.

TABLE II-2

Determinants of Evolutionary Performance (Technological Yield)

Dependent Variable	Technological Yield
Intercept	–396 (375)
Time	0.20 (0.19)
DRAM	14.6*** (1.61)
Project resources (PY)	–0.0015 (0.0051)
Experimental capacity (wafers/week)	0.00078 (0.0012)
Experimental iteration time (weeks)	–0.56* (0.28)
No previous project experience	–9.6** (3.9)
Research experience	–0.28 (1.30)
F-test	29.58***
R-squared	0.89
N	29

The dependent variable is the logarithm of the transistor density. Time is a continuous variable corresponding to the time of process introduction. The DRAM dummy is *1* if the project was aimed at the DRAM segment, and *0* otherwise. A * means significance at 10 percent level, ** significance at 5 percent level, and *** significance at 1 percent level.

Bibliography

Abernathy, W. J. *The Productivity Dilemma.* Baltimore: Johns Hopkins University Press, 1978.

Abernathy, W. J., and K. B. Clark. "Innovation: Mapping the Winds of Creative Destruction." *Research Policy* 14, no. 1 (1985): 3–22.

Abernathy, W. J., and J. M. Utterback. "Patterns of Industrial Innovation." *Technology Review* 80, no. 7 (1978): 40–47.

Adler, Paul S., and K. B. Clark. "Behind the Learning Curve: A Sketch of the Learning Process." *Management Science* 37, no. 3 (1991): 267–281.

Alexander, C. *Notes on the Synthesis of Form.* Cambridge, MA: Harvard University Press, 1964.

Allen, T. J. "Studies of the Problem-Solving Process in Engineering Design." *IEEE Transactions on Engineering Management* EM-13, no. 2 (1966): 72–83.

Allen, T. J. *Managing the Flow of Technology.* Cambridge, MA: MIT Press, 1977.

Allen, T. J. "Organizational Structures, Information Technology and R&D Productivity." *IEEE Transactions on Engineering Management* EM-33, no. 4 (1986): 212–217.

Allen, T. J., and O. Hauptman. "The Influence of Communication Technologies on Organizational Structure." *Communication Research* 14, no. 5 (1987): 575–578.

Allen, T. J., D. M. S. Lee, and M. L. Tushman. "Technology Transfer as a Function of Position in the Spectrum from Research Through Development to Technical Services." *Academy of Management Journal* 22, no. 4 (1980): 694–708.

Allen, T. J., M. L. Tushman, and D. M. S. Lee. "R&D Performance as a Function of Internal Communication, Project Management, and the Nature of Work." *IEEE Transactions on Engineering Management* EM-27, no. 1 (1980): 2–12.

Anderson, P., and M. L. Tushman. "Technological Discontinuities and Dominant Designs: A Cyclical Model of Technological Change." *Administrative Science Quarterly* 35 (1990): 604–633.

Argyris, C. *On Organizational Learning.* Cambridge, MA: Blackwell Business, 1995.

Argyris, C., and D. Schon. *Organizational Learning.* Reading, MA: Addison-Wesley, 1978.

Arrow, K. "Economic Welfare and the Allocation of Resources of Invention." In *The Rate and Direction of Inventive Activity: Economic and Social Factors*, edited by R. Nelson. Princeton, NJ: Princeton University Press, 1962.

Arrow, Kenneth. *The Limits of Organization*. New York: Norton, 1974.

Baldwin, C. Y., and K. B. Clark. "Modularity and Real Options." Working paper 93-026, Harvard Business School, Boston, MA, 1994.

Bowen, H. K., Kim B. Clark, Charles Holloway, and Steven C. Wheelwright, eds. *Vision and Capability: High Performance Product Development in the 1990s*. New York: Oxford University Press, 1993.

Bowen, H. K., K. B. Clark, C. A. Holloway, and S. C. Wheelwright. *The Perpetual Enterprise Machine*. New York: Oxford University Press, 1994.

Burgelman, R. A. "Fading Memories: A Process Theory of Strategic Business Exit in Dynamic Environments." *Administrative Sciences Quarterly* 39, no. 1 (1994):24–56.

Burns, T., and G. M. Stalker. *The Management of Innovation*. London: Tavistock Publications, 1961.

Chandler, A. D., Jr. *The Visible Hand: The Managerial Revolution in American Business*. Cambridge, MA: The Belknap Press of Harvard University Press, 1977.

Chandler, A. D., Jr. *Scale and Scope: The Dynamics of Industrial Capitalism*. Cambridge, MA: The Belknap Press of Harvard University Press, 1990.

Chandler, A. D., Jr. "Corporate Strategy, Structure, and Control Methods in the United States during the Twentieth Century." *Industrial and Corporate Change* 1, no. 2 (1992): 263–284.

Christensen, C. "The Innovator's Challenge: Understanding the Influence of Market Environment on Process Technology Development in the Rigid Disk Drive Industry." Ph.D. diss., Harvard Business School, 1992.

Christensen, C. "The Drivers of Vertical Disintegration." Working paper 96-008, Harvard Business School, Boston, MA, 1994.

Christensen, C. *The Innovator's Dilemma: When New Technologies Cause Great Firms to Fail*. Cambridge, MA: Harvard Business School Press, 1997.

Christensen, C. and Rosenbloom, R. "Explaining the Attacker's Advantange: Technological Paradigms, Organizational Dynamics, and the Value Network." *Research Policy* 24, no. 2 (1995): 233–257.

Clark, K. B. "The Interaction of Design Hierarchies and Market Concepts in Technological Evolution." *Research Policy* 14, no. 5 (1985): 235–251.

Clark, K. B., W. B. Chew, and T. Fujimoto. "Product Development in the World Auto Industry: Comments and Discussion." *Brookings Papers on Economic Activity* 3 (1987): 729–771.

Clark, K. B., and T. Fujimoto. "Lead Time in Automobile Product Development: Explaining the Japanese Advantage." *Journal of Engineering and Technology Management* 6, no. 1 (1989): 25–58.

Clark, K. B., and T. Fujimoto. "Overlapping Problem Solving in Product Development." In *Managing International Manufacturing*, edited by K. Ferdows; 127-152. Amsterdam: North-Holland, 1989.

Clark, K. B., and T. Fujimoto. "The Power of Product Integrity." *Harvard Business Review* 68, no. 6 (1990): 107–118.

Clark, K. B., and T. Fujimoto. *Product Development Performance*. Boston, MA: Harvard Business School Press, 1991.

Clausing, D. *Total Quality Development: A Step by Step Guide to World Class Concurrent Engineering*. New York: ASME Press, 1993.

Cogan, G. W., and Burgelman, R. A. "Intel Corporation (A): The DRAM Decision." Case PS–BP–256. California: Stanford Business School, 1990.

Cogan, G. W., and Burgelman, R. A. "Intel Corporation (C): Strategy for the 1990s." Case PS–BP–256C. California: Stanford Business School, 1991.

Cohen, H., S. Keller, and D. Streeter. "The Transfer of Technology from Research to Development." *Research Management* 22, no. 3 (1979): 11–17.

Cohen, W. M., and D. A. Levinthal. "Absorptive Capacity: A New Perspective on Learning and Innovation." *Administrative Sciences Quarterly* 35, no. 1 (1990): 128–152.

Collins, H. M. "Tacit Knowledge in Scientific Networks." In *Science in Context: Readings in the Sociology of Science*, edited by B. Barnes and D. Edge, Cambridge, MA: MIT Press, 1982.

Cooper, R. G. "A Process Model for Industrial New Product Development." *IEEE Transactions on Engineering Management* EM-30, no. 1 (1983): 2–11.

Cooper, R. G. "Stage-Gate Systems: A New Tool for Managing New Products." *Business Horizons* 33, no. 3 (1990): 44–54.

Cusumano, M., and R. Selby. *Microsoft Secrets*. New York: Free Press, 1995.

Cusumano, M. A. "Shifting Economies: From Craft Production to Flexible Systems and Software Factories." *Research Policy* 21, no. 5 (1992): 453–480.

Cusumano, M. A., and K. Nobeoka. "Strategy, Structure and Performance in Product Development: Observations from the Auto Industry." *Research Policy* 21, no. 3 (1992): 265–293.

Daft, R. L., and K. E. Weick. "Toward a Model of Organizations as Interpretation Systems." *Academy of Management Review* 9, no. 2 (1984): 284–295.

Davari, B., and R. H. Dennard. "CMOS Scaling for High Performance and Low Power—the Next Ten Years." Proceedings of *IEEE* 83, no. 4 (1995): 595–606.

Dewar, R. D., and J. E. Dutton. "The Adoption of Radical and Incremental Innovations: An Empirical Analysis." *Management Science* 32, no. 11 (1986): 1422–1433.

Dohya, A., T. Watari, and H. Nishimori. "Packaging Technology for the NEC SX-3/SX-X Supercomputer." *IEEE Proceedings*, 1990.

Dosi, G., and L. Marengo. "Some Elements of an Evolutionary Theory of Organizational Competences." In *Evolutionary Concepts in Contemporary Economics*, edited by R. W. England. Ann Arbor: University of Michigan Press, 1993.

Dosi, G., D. J. Teece, and S. Winter. "Toward a Theory of Corporate Coherence." In *Technology and Enterprise in a Historic Perspective*, edited by G. Dosi, R. Gianetti, and A. Toninelli, 186–211. Oxford: Clarendon Press, 1992.

Eisenhardt, K. M., and B. N. Tabrizi. "Accelerating Adaptive Processes: Product Innovation in the Global Computer Industry." *Administrative Sciences Quarterly* 40, no. 1 (1995): 84–110.

Ettlie, J. E., W. P. Bridges, and R. D. O'Keefe. "Organizational Strategy and Structural Differences for Radical vs. Incremental Innovation." *Management Science* 30, no. 6 (1984): 682–695.

Flaherty, M. T. "Manufacturing and Firm Performance in Technology Intensive Industries: U.S. and Japanese DRAM Experience." Working paper 92-070, Harvard Business School, Boston, MA, 1992.

Frischmuth, D. S., and T. J. Allen. "A Model for the Description and Evaluation of Technical Problem Solving." *IEEE Transactions on Engineering Management.* EM-16, no. 2 (1969): 58–64.

Fujimoto, T. "Organizations for Effective Product Development: The Case of the Global Automobile Industry." D.B.A. diss., Harvard Business School, 1989.

Fujimoto, T., M. Iansiti, and K. B. Clark. "External Integration in Product Development." In *Managing Product Development*, edited by T. Nishiguchi. New York: Oxford University Press, 1996.

Galbraith, J. R. *Designing Complex Organizations.* Reading, MA: Addison-Wesley, 1973.

Garvin, D. "Quality Problems, Policies, and Attitudes in the United States and Japan: An Exploratory Study." *Academy of Management Journal* 29, no. 4 (1996): 653–673.

Gibbons, M., and R. D. Johnston. "The Role of Science in Technological Innovation." *Research Policy* 3, no. 3 (1974): 220–242.

Graham, B. K., and Burgelman, R. A. "Intel Corporation (B): Implementing the DRAM Decision." Case PS–BP–256B. California: Stanford Business School, 1991.

Griliches, Z., ed. *R&D, Patents, and Productivity.* Chicago: University of Chicago Press, 1984.

Hannan, M. T., and J. Freeman. "Structural Inertia and Organizational Change." *American Sociological Review* 49 (1984): 149–164.

Hauptman, O., and S. L. Pope. "The Process of Applied Technology Forecasting." *Technological Forecasting and Social Change* 42, no. 2 (1992): 193–211.

Hayes, R. H., S. C. Wheelwright, and K. B. Clark. *Dynamic Manufacturing.* New York: Free Press, 1988.

Henderson, R. "The Evolution of Integrative Capability: Innovation in Cardiovascular Drug Discovery." *Industrial and Corporate Change* 3, no. 3 (1994): 607–630.

Henderson, R. "Of Lifecycles Real and Imaginary: The Unexpectedly Long Old Age of Optical Lithography." *Research Policy* 24, no. 4 (1995): 631–643.

Henderson, R., and K. B. Clark. "Architectural Innovation: The Reconfiguration of Existing Product Technologies and the Failure of Established Firms." *Administrative Sciences Quarterly* 35, no. 1 (1990): 9–30.

Henderson, R., and I. Cockburn. "Scale, Scope and Spillovers: The Determinants of Research Productivity in Drug Discovery." *Rand Journal of Economics* 27, no. 1 (1996): 32–59.

Hounshell, D. A., and J. K. Smith. *Science and Corporate Strategy: Dupont R&D, 1902–1980.* New York: Cambridge University Press, 1988.

Iansiti, M. "Technology Integration: Exploring the Interaction Between Applied Science and Product Development." Working paper 92-026, Harvard Business School, Boston, MA, 1991; rev. 1992.

Iansiti, M. "Science-Based Product Development: An Empirical Study of the Mainframe Computer Industry." *Production and Operations Management* 4, no. 4 (1995): 335–359.

Iansiti, M. "Technology Development and Integration: An Empirical Study of the Interaction Between Applied Science and Product Development." *IEEE Transactions on Engineering Management* 42, no. 2 (1995): 259–269.

Iansiti, M. "Technology Integration: Managing the Interaction Between Applied Science and Product Development." *Research Policy* 24, no. 4 (1995): 521–524.

Iansiti, M. "From Technological Potential to Product Performance: An Empirical Analysis." Working paper 96-007, Harvard Business School, Boston, MA, 1995.

Iansiti, M. "Shooting the Rapids: Managing Product Development in Turbulent Times." *California Management Review* 38, no. 1 (1995): 37–58.

Iansiti, M. "Managing Technological Transitions: Empirical Evidence from the Semiconductor Industry." Working paper 96-046, Harvard Business School, Boston, MA, 1997.

Iansiti, M., and K. B. Clark. "Integration and Dynamic Capability: Evidence from Product Development in Automobiles and Mainframe Computers." *Industrial and Corporate Change* 3, no. 3 (1994): 557–605.

Iansiti, M., and T. Khanna. "Technological Evolution, System Architecture, and the Obsolescence of Firm Capabilities." *Industrial and Corporate Change* 4, no. 2 (1995): 333–361.

Iansiti, M., and A. MacCormack. "Developing Products on Internet Time." Working paper 97-027, Harvard Business School, Boston, MA, 1997.

Ikari, Y. *Nissan Ishiki Daikakume (Great Cultural Revolution of Nissan).* Tokyo: Diamond, 1987. [Japanese language.]

Katz, R. "The Effects of Group Longevity on Project Communication and Performance." *Administrative Sciences Quarterly* 27, no. 1 (1982): 81–104.

Katz, R., and T. J. Allen. "Project Performance and the Locus of Influence in the R&D Matrix." *Academy of Management Journal* 28, no. 1 (1985): 67–87.

Keller, R. T. "Predictors of the Performance of Project Groups in R&D Organizations." *Academy of Management Journal* 29, no. 4 (1986): 715–726.

Khanna, T., and M. Iansiti. "Firm Asymmetries and Sequential R&D: Theory and Evidence from the Mainframe Computer Industry." Working paper 94-006, Harvard Business School, Boston, MA, 1993.

Kiesler, S., and L. Sproull. "Managerial Response to Changing Environments: Perspectives on Problem Sensing from Social Cognition." *Administrative Science Quarterly* 27, no. 4 (1982): 548–570.

Langrish, J. "Technology Transfer: Some British Data." *R&D Management* 1, no. 3 (1971): 133–136.

Lant, T., and S. Mezias. "An Organizational Learning Model of Convergence and Reorientation." *Organization Science* 3, no. 1 (1992): 47–71.

Lawrence, P. R., and J. W. Lorsch. *Organization and Environment.* Boston, MA: Harvard Business School Press, 1967.

Lazonic, W. *Competitive Advantage on the Shop Floor.* Cambridge, MA: Harvard University Press, 1990.

Leonard-Barton, D. "Implementation as Mutual Adaptation of Technology and Organization." *Research Policy* 17, no. 5 (1988): 251–267.

Leonard-Barton, D. "Core Capabilities and Core Rigidities: A Paradox in Managing New Product Development." *Strategic Management Journal* 13 (1992): 111–125.

Levitt, B., and J. March. "Organizational Learning." *Stanford University Annual Review* 14 (1988): 319–340.

March, J. C., and J. G. March. "Almost Random Careers: The Wisconsin School Superintendency, 1940–1972." *Administrative Science Quarterly* 22 (1977): 377–409.

March, J. G., and H. A. Simon. *Organizations.* New York: Wiley, 1958.

Marple, D. L. "The Decisions of Engineering Design." *IEEE Transactions of Engineering Management* 2, no. 1 (1961): 55–71.

Marquis, D. G., and D. L. Straight. "Organizational Factors in Project Performance." Working paper 133-65, MIT Sloan School of Management, Boston, MA, 1965.

McDonough, E. F., III, and G. Barczak. "The Effects of Cognitive Problem-Solving Orientation and Technological Familiarity on Faster New Product Development." *Journal of Product Innovation Management* 9, no. 1 (1992): 44–52.

McGrath, R. G., I. C. MacMillan, and M. L. Tushman. "The Role of Executive Team Actions in Shaping Dominant Designs: Towards the Strategic Shaping of Technological Progress." *Strategic Management Journal* 13 (1992): 137–161.

Meyer, C. *Fast Cycle Time.* New York: Free Press, 1993.

Mintzberg, H., D. Raisinghani, and A. Theoret. "The Structure of 'Unstructured' Decision Processes." *Administrative Science Quarterly* 21, no. 2 (1976): 246–275.

Moch, M., and E. V. Morse. "Size, Centralization, and Adoption." *American Sociological Review* 42 (1977): 716–725.

Nelson, R., and S. Winter. *An Evolutionary Theory of Economic Change.* Cambridge, MA: Harvard University Press, 1982.

O'Reilly C., III, D. Calwell, and W. Barnett. "Work Group Demography, Social Intergration, and Turnover." *Administrative Science Quarterly* 34, no. 1 (1989): 21–37.

Penrose, E. *The Theory of the Growth of the Firm.* London: Basil Blackwell, 1959.

Perrow, C. "A Framework for Comparative Organizational Analysis." *American Sociological Review* 32 (1967): 194–208.

Pisano, G. "Learning Before Doing in the Development of New Process Technology." *Research Policy* 25, no. 7 (1996): 1097–1119.

Pisano, G. *The Development Factory: Unlocking the Potential of Process Innovation.* Boston: Harvard Business School Press, 1997.

Polanyi, M. *Personal Knowledge: Towards a Post-Critical Philosophy.* Chicago: University of Chicago Press, 1958.

Porter, M. E. *Competitive Strategy: Techniques for Analyzing Industries and Competitors.* New York: Free Press, 1980.

Prahalad, C. K., and Hamel, G. "The Core Competence of the Corporation." *Harvard Business Review* 68, no. 3 (1990): 79–91.

Pugh, E. M. *Memories That Shaped an Industry.* Cambridge, MA: MIT Press, 1984.

Rappaport, A., and S. Halevi. "The Computerless Computer Company." *Harvard Business Review* 69, no. 4 (1991): 69–80.

Rosenberg, N. *Inside the Black Box: Technology and Economics.* New York: Cambridge University Press, 1982.

Rosenbloom, R. S., and W. J. Spencer, eds. *Engines of Innovation: U.S. Industrial Research at the End of an Era.* Boston, MA: Harvard Business School Press, 1996.

Saviotti, P. P., and J. S. Metcalfe. "A Theoretical Approach to the Construction of Technological Output Indicators." *Research Policy* 13, no. 3 (1984): 141–151.

Schrader, S., W. M. Riggs, and R. P. Smith. "Choice over Uncertainty and Ambiguity in Technical Problem Solving." *Journal of Engineering and Technology Management* 10, no. 1, 2 (1992): 73–99.

Schumpeter, J. *The Theory of Economic Development.* Cambridge, MA: Harvard University Press, 1934.

Sherwin, E. W., and R. S. Isenson. "Project Hindsight." *Science* 156 (1967): 1571–1577.

Simon, H. A. "Rationality as Process and as Product of Thought." *American Economic Review* 69, no. 4 (1978): 1–16.

Smith, G. P. G., and D. G. Reinertsen. *Developing Products in Half the Time.* New York: Van Nostrand, 1991.

Stalk, G., Jr., "Time—The Next Source of Competitive Advantage." *Harvard Business Review* 66, no. 4 (1988): 41–51.

Teece, D. J. "Towards an Economic Theory of the Multiproduct Firm." *Journal of Economic Behavior and Organization* 3, no. 1 (1982): 39–63.

Teece, D. J., G. Pisano, and A. Shuen. "Dynamic Capabilities and Strategic Management." *Strategic Management Journal* (1997): In Press.

Thomke, S. "The Economics of Experimentation in the Design of New Products and Processes." Ph.D. diss., MIT Sloan School of Management, 1995.

Thomke, S. "Managing Experimentation in the Design of New Products and Processes." Working paper 96-037, Harvard Business School, Boston, MA, 1996.

Thomke, S. H., E. S. Von Hippel, and R. Framke. "Modes of Experimentation: An Innovation Process Variable." Working paper 97-057, Harvard Business School, Boston, MA, 1997.

Thompson, J. *Organizations in Action.* New York: McGraw-Hill, 1967.

Tummala, R. R., and E. J. Rymaszewski, eds. *Microelectronics Packaging Handbook.* New York: Van Nostrand, 1988.

Tushman, M. L., and P. Anderson. "Technological Discontinuities and Organizational Environments." *Administrative Science Quarterly* 31, no. 3 (1986): 439–465.

Tushman, M. L., and D. Nadler. "Communication and Technical Roles in R&D Laboratories: An Information Processing Approach." *Management Science, Special Studies: R&D Management* 15 (1980): 91–112.

Tushman, M. L., and C. A. O'Reilly, III. *Winning Through Innovation.* Boston, MA: Harvard Business School Press, 1997.

Tushman, M. L., and E. Romanelli. "Organizational Evolution: A Metamorphosis Model of Convergence and Reorientation." *Research in Organizational Behavior* 7 (1985): 171–222.

Tushman, M. L., and L. Rosenkopf. "Organizational Determinants of Technological Change: Toward a Sociology of Technological Evolution." *Research in Organizational Behavior* 14 (1992): 311–347.

Tyre, M. J. "Managing Innovation on the Factory Floor." *Technology Review* 94, no. 7 (1991): 59–65.

Tyre, M. J. "Managing the Introduction of New Process Technology: International Differences in a Multi-Plant Network." *Research Policy* 20, no. 1 (1991): 1–21.

Tyre, M. J., and O. Hauptman. "Effectiveness of Organizational Response Mechanisms to Technological Change in the Production Process." *Organization Science* 3, no. 3 (1992): 301–320.

Tyre, M. J., and W. J. Orlikowski. "Windows of Opportunity: Temporal Patterns of Technological Adaptation in Organizations." *Organization Science* 5, no. 1 (1994): 98–118.

Tyre, M. J., and E. Von Hippel. "The Situated Nature of Adaptive Learning in Organizations." *Organization Science* 8, no. 1 (1997): 71–83.

Ulrich, K. T., and S. D. Eppinger. *Product Design and Development.* New York: McGraw-Hill, 1994.

Uttal, B. "Speeding Ideas to Market." *Fortune* (2 March 1987): 62–66.

Utterback, J. M. *Mastering the Dynamics of Innovation: How Companies Can Seize Opportunities in the Face of Technological Change.* Boston, MA: Harvard Business School Press, 1994.

Van de Ven, A. "Central Problems in the Management of Innovation." *Management Science* 32, no. 5 (1986): 590–607.

Van de Ven, A. H., and R. Drazin. "The Concept of Fit in Contingency Theory." *Research in Organizational Behavior* 7 (1985): 333–365.

Venkatraman, N. "The Concept of Fit in Strategy Research: Towards Verbal and Statistical Correspondence." *Academy of Management Review* 14, no. 3 (1989): 423–444.

Virany, B., M. Tushman, and E. Romanelli. "Executive Succession and Organization Outcomes in Turbulent Environments: An Organization Learning Approach." *Organization Science* 3, no. 1 (1992): 72–91.

Von Hippel, E. "Task Partitioning: An Innovation Process Variable." *Research Policy* 19, no. 5 (1990): 407–418.

Von Hippel, E. "The Impact of 'Sticky Data' on Innovation and Problem Solving." *Management Science* 40, no. 4 (1994): 429–439.

Von Hippel, E., and M. Tyre. "How Learning Is Done: Problem Identification in Novel Process Equipment." *Research Policy* 24, no. 1 (1995): 1–12.

Wagner, W. G., J. Pfeffer, and C. A. O'Reilly, III. "Organizational Demography and Turnover in Top-Management Groups." *Administrative Science Quarterly* 29, no. 1 (1984): 74–92.

Weick, Karl E. *The Social Psychology of Organizing.* New York: McGraw-Hill, 1979.

Wernerfelt, B. "A Resource-Based View of the Firm." *Strategic Management Journal* 5, no. 2 (1984): 171–180.

West, J. "Institutional Diversity and Modes of Organization for Advanced Technology Development: Evidence from the Semiconductor Industry." Ph.D. diss., Harvard Business School, 1996.

Wheelwright, S. C., and K. B. Clark, *Revolutionizing Product Development.* New York: Free Press, 1992.

Woodward, J. *Industrial Organization.* London: Oxford University Press, 1965.

Index

ACOS mainframes, 130
Alpha, 82
Altmaier, Rich, 140
ambidextrous management, 20
AMD, 154
ASIC hardware, problem-solving, 105–106
AT&T Bell Laboratories, 150, 209
Atlas project, 188

Bahr, Rick, 140
beta testing
  Internet software development, 196–198
  knowledge development and, 202–203
  Microsoft Explorer, 199–200
  Netscape Navigator, 188–190
  See also feedback; testing
block diagrams, 70
bring-up stage
  problem-solving in, 103
  in revolutionary projects, 142
browser software, 187–190. *See also* Netscape Navigator
buckling problem, mainframes, 100–102
bugs, software development and, 179–180, 197–198
Burgelman, Robert, 154
bus architecture, 134, 136t

career issues, 71, 165–167
Cashmere project, 177–181
centralization, semiconductor industry, 168
ceramic buckling, problem-solving, 100–102
ceramic substrate processing, 100
Challenge project (Silicon Graphics), 66–71, 134–138, 139
  knowledge application, 67–69
change. *See* evolutionary change; revolutionary change; technological change
client engineering, 188
CMOS DRAMs, 158
code
  development and debugging, 179–180, 197–198
  See also software development
Code Complete milestone, 179
communication, organizational, 69
comparison indicators, 79–82
competitiveness, 51, 59, 154
complexity
  development and, 13
  empirical setting and, 33
  experimentation and, 22
  innovation management and, 12
  novelty and, 2, 8
  problem framing and, 108
  problem-solving and, 106–107
  product subsystem interactions, 78–79
  research and development and, 211
  in software development, 174–177, 185
  technical choice and, 2, 71–72
computer industry
  predicted demise of, 149
  recovery of, 149–165
concept design and development, software, 184–187, 201, 203–204
concept-freeze milestone, 184, 185
concept lead time
  defined, 39–40, 41t
  project performance and, 38t, 39–40, 41t
  in software development, 185
concurrent engineering, 186–187
context-independent knowledge domains. *See* fundamental knowledge domains

context-independent problem-solving breadth, 110, 112
  organizational process and, 115
context-specific knowledge
  defined, 11
  matching fundamental knowledge with, 76
  merging fundamental knowledge with, 102
  obsolescence of, in revolutionary projects, 143
  problem-solving and, 107, 110
  technology choice and, 114
context-specific problem-solving breadth, 110, 112
  organizational process, and, 115–117
  product performance and, 114
"Copy Exactly" technology transfer, 160, 169
customers
  feedback from, 182
  integration of, 8
  technological choice and, 8–9
  See also beta testing
cycle time, 79
Cyrix, 154

Dash project, 138
debugging
  Internet software, 197–198
  software code, 179–180
decision-making approach
  project performance and, 66
  See also problem-solving
deep ultraviolet lithography (DUV), 56
delay time, interconnect performance and, 218
detailed design, in software development, 185–187
detrended gate density, 57
development capability index, 47
development lead time
  defined, 39, 41t
  performance and, 50t
  problem-solving breadth and, 112
  project performance and, 38t, 40–42, 52t–54t, 66
  project resources and, 42
  in software development, 185
  technology integration and, 52t–54t
development process
  product performance and, 93–94
  project performance and, 42–47
  software, 175–187
  system-focused, 95
  See also module development; process development; research and development (R&D); software development
development project manager, internal integration and, 44t
development projects
  goals of, 13
  organizational structure and, 14
  technology integration and, 21
development routines, inertia and, 16–17
development speed, variables affecting, 42
dielectric constant, 218, 220
Digital Equipment Corporation (DEC)
  Alpha processor, 77, 79
  competitiveness, 154
distributed shared memory (DSM) architecture, 134–139
  servers, 71
  strategic importance of, 214
domain-specific knowledge
  defined, 10
  problem solving and, 114–115
  technological potential and, 89
  technology integration and, 119
domains. *See* knowledge domains
DRAMs
  CMOS, 158
  competition, 51
  development, 51, 55, 56–59
  empirical setting, 33
  industry, 151–156
  Intel, 155–159
  investment, 51
  NMOS, 158
  performance averages among firms, 58t
  technology integration, 155–159
DUV, 56

dynamic random-access memory. *See* DRAMs

education
of personnel, 163
problem-solving and, 107
See also experience
EEPROMs, Intel, 155, 158
efficiency, project resources and, 88
empirical environments, 32–33
advanced computer processors, 84–85
field work design, 28–32
mainframes, 37
semiconductor process technology, 55–56
employment
lifetime, 167–168, 212–213
new hires, 65
Ph.D.-level hires, 166–167
project team turnover, 64–65, 144, 166, 180
EPROMs, Intel, 155, 158
equipment suppliers
reliance on, 159–160
See also external technologies
Europe, 49, 53t
evolutionary change
defined, 123
as response to technological change, 122–123
technological yield as measure of, 123–124
evolutionary projects
characteristics of, 141–143, 144
defined, 123–124
effectiveness of, 143–144
experience in, 133
experimentation in, 133, 145
inertia and, 145
mainframes, 129–130
model for managing, 126–128
regression analysis, 223
revolutionary foundations of, 129–132
semiconductors, 124–126, 169–170
supercomputers, 128–134
variables associated with, 128t
Excite, 190
expectations, technology integration failure and, 180
experience, 61
in evolutionary projects, 127–128, 133, 144
excessive, 24
Internet software, 204, 205
of personnel, 163–164, 165t
personnel retention, 212–213
project performance and, 70
semiconductor industry, 166–167
system knowledge and, 24
turnover and, 180
See also project experience; research experience
experimentation
in evolutionary projects, 127–128, 133, 143–144
facility design, 23
Internet software, 192, 204–205
iteration time, 23, 61, 64
knowledge generation and, 22
massive, 93, 140, 142, 145
process development and, 156, 160
project performance and, 66–67
in revolutionary projects, 127–128, 141–142, 145
selective, in evolutionary projects, 145
semiconductor industry, 161–163, 167–168, 169–170
set-up, representativeness of, 23
technological potential and, 93
technology choice and, 64, 161
technology integration driven by, 212
technology integration effectiveness and, 214
throughput time, 163
workstations, 66–67
experimentation capacity, 61
defined, 22–23
in evolutionary projects, 144
process performance and, 64
results of limited, 156
semiconductor industry, 162
external technologies
integration of, 207–208
need for, 210–211
technology integration capability and, 165–166

external testing
Internet software, 202, 203
Silicon Graphics Challenge project, 70
slow release method, 191

feasibility studies
supercomputer packaging, 130–131
technology integration process and, 214
feedback
beta testing, 188–190
external testing, 70, 191, 202, 203
internal testing, 199, 202–203
intranet, 190
newsgroups, 189
See also beta testing; testing
field work design, 28–32
first-loop learning processes, 145
flexibility
in Internet software development, 190–191, 205–206
knowledge generation and, 201–202
in software development, 183–187
workstations and server requirements, 66
focus
narrow, in research projects, 13
semiconductor industry, 168–169
frames of reference, in problem-solving, 107–108
fundamental knowledge domains
context-specific knowledge and, 102
defined, 76
in problem-solving, 110
technological potential and, 89

Galles, Mike, 140
gate density, 57, 152
competitiveness and, 59
detrended, 57
evolutionary/revolutionary nature, 124
performance determinants, 63t
technological potential and realized performance, 85t
Gates, Bill, 175, 176, 177–178, 180, 189, 199

Hennessy, John, 138
Hewlett-Packard, 154
high-volume logic development, 56
Hitachi, 151

IBM
DRAMs, 151
industry problems, 149
industry recovery, 149
Semiconductor Research and Development Center (SRDC), 126, 213
semiconductors, 154
technological potential values, 124–126
technology integration, 212–214
T. J. Watson Research Center, 125–126, 150, 209
implementation, in software development, 184–187, 201, 203–204
industrial research, technology integration needs, 211–212
inertia
development routines and, 16–17
during periods of technological change, 121
evolutionary projects and, 145
ineffective technology integration and, 25
managing, 18
organizational change and, 18
revolutionary projects and, 145
in traditional research and development, 15–18
"Infinite Defects," 179
information technology, 69
integrated circuits
evolutionary/revolutionary nature, 124
performance determinants, 63t
integrated products, 75, 79
Intel
DRAMs, 151
experimentation, 161–162
industry problems, 149
industry recovery, 149
production facility investment, 2–3
semiconductors, 154–161
technology integration, 63, 212, 213

interactions, knowledge of, 21
interconnect system
  product performance, 220–221
  role of, 217
  technological yield, 217, 221
internal integration
  defined, 43
  index, project performance and, 43–45
internal testing
  Internet software, 202–203
  using intranets, 199
International Business Machines. *See* IBM
Internet software
  beta testing, 188–190, 196–198, 199–200, 202–203
  browsers, 187–190
  development trends, 173–175
  experience, 204
  experimentation, 204–205
  flexibility needs, 190–191, 205
  integration, 203–204
  knowledge generation, 201–203
  market and, 206–208, 211
  Microsoft Explorer, 199–201
  My Yahoo!, 192–194
  NetDynamics, 194–199
  Netscape Navigator, 187–190
  novelty and complexity in, 174–175
  prototypes, 204–205
  responsiveness issues, 205–208
  search programs, 174
  soft release, 203
  technology integration process, 174–177, 206–208, 213
  time scales, 174
  Yahoo!, 190–194
  See also software development
intranets
  feedback through, 190, 199–200
  software, 188
intuition, 168
IRIS, 77–78, 79, 81

Japan
  lifetime employment, 212–213
  semiconductors, 151, 152–153, 165–170
  technology integration capability index, 48–49
  technology integration/project performance relationship, 52t
Java, 195–196, 214

Klammer, Franz, 82
knowledge
  domain-specific, 10, 89, 114–115, 119
  of interactions, 21
  internal, 90
  problem-solving and, 99
  reflected in products, 9–10
  research and development process effectiveness and, 10
  scientific, 210–211
  system, 24
  technology integration and, 21
knowledge application
  defined, 24–25
  performance and, 67–69
  project performance and, 37, 62–64
  semiconductor industry, 168–169
  technology integration capability and, 47t, 165t
  variables, 61t
knowledge domains
  context-specific, 11, 76, 102, 107, 110, 114, 143
  deep specialization, 13
  defined, 10
  domain-specific, 10, 89, 102, 114–115, 119
  fundamental (context-independent), 21, 76, 89, 102, 110
  integration of, 11
  interactions among, 12, 89, 114
  interconnect performance and, 220
  merging fundamental and specific, 102
  in problem-solving, 110, 114
  research projects and, 12–13
  technological yield and, 89
  unpredictability of, 12
  See also domain-specific knowledge
knowledge generation, 21–23
  flexibility of software development and, 201–203
  performance and, 69–70

knowledge generation *(continued)*
project performance and, 36, 64
in revolutionary projects, 141–142, 145
in semiconductor industry, 167–168
technology integration capability and, 48t
technology selection and, 21–22
through experimentation, 161–163
variables, 61t
knowledge retention
defined, 23–24
organizational practice and, 212–213
performance and, 64–65
project performance and, 36, 70–71
in semiconductor industry, 166–167
technology integration capability and, 49t, 164t
variables, 62t
Korea, 151, 165–170

leadership
in software development, 177, 180–181
See also management
lead time
concept, 38, 39–40, 41, 185
development, 31, 38, 39, 40–42, 50, 52t–54t, 66, 112, 185
in software development, 185
total, 38t, 40, 41, 87–88, 185
variables, project performance and, 38t, 39–42
learning processes
first-loop, 145
second-loop, 145
Lego project, 18, 19–20, 71, 134, 138–141
problem-solving in, 103–106
Lenoski, Dan, 138, 139
lifetime employment, 167, 212–213
localization lead, role of, 177
logic environment
circuit development, 57
improvements, 153–155
logic gates, interconnect system and, 217
Lauden, Jim, 138, 139
Lycos, 190

McCracken, Ed, 215
mainframes, 35–51
buckling problem, 100–102
empirical approach to product performance, 84–85
empirical environment, 37
evolutionary change, 129–130
performance characteristics, 81–82
problem-solving, 100–106, 110–119
processor modules, comparison indicators, 79–82
project performance, 37–42, 47–51
relating process to performance, 42–46
technology integration, 47–51
technology integration capability assessment, 46–47
management
ambidextrous, 20
complexity and, 12
of evolution and revolution, 126–128
novelty and complexity and, 12
program manager, 177, 178, 180–183
project, 93–94
technological change and, 18–20
technology choice and, 20
technology integration and, 18, 121, 145–146, 182
See also leadership
manufacturing capacity, semiconductors, 151–152
marketplace
Internet development and, 206–208, 211
product introduction to, technological yield and, 95
workstation and server requirements, 66
Metacrawler, 190
microprocessors
Intel, 155
logic project improvements, 153–155
performance averages among firms, 59t
process development, 158–159
technological potential, 124–126
technological yield, 124–126
Microsoft
development process improvements, 181–183

early development, 175–177
experience, 205
experimentation, 205
industry problems, 149
industry recovery, 149
intranet, 199–200
market preferences and, 207
Office Suite, 178
on-line environment, 173
software development process, 199–201
technology integration process, 213
Windows 95, 3
Windows NT, 194
Word for Windows 1.0, 177–181
Microsoft Internet Explorer, 189, 199–201
beta testing, 199–200
market preferences and, 207
MIPS, 126
models
physical, 69–70
research and development, 12–15
subsystem-level, 131
See also prototypes; simulations
module development
mainframes, project performance, 37–42, 47–51
relating process to performance, 42–46
technology integration, 47–51
technology integration capability assessment, 46–47
Mosaic, 188
Motorola, 154
multichip module technology, project performance, 37
My Yahoo!
experience, 204–205
experimentation, 204
integration, 203–204
knowledge development, 202–203
prototypes, 204
software development, 192–194

Nakahira, David, 138, 139
NEC, 128–134, 151
NetDynamics, 194–199, 201
experimentation, 204
integration, 204
knowledge development, 202
NetDynamics 2.0, 195–199
prototypes, 196, 204
Runtime, 196–197
scripting language choice, 195–196
Spider 1.5, 194
Studio, 196–197
technology integration process, 214
Netscape Communications, 173, 187–190
beta testing, 202
experimentation, 214
knowledge development, 202
technology integration process, 214
Netscape Navigator, 187–190, 199
beta testing, 188–190
experimentation, 204–205
Navigator 2.0, 188
Navigator 3.0, 188–190
prototypes, 188–190, 204–205
new hires
performance and, 65
Ph.D. level, 166–167
newsgroups, feedback through, 189
NexGen, 154
NMOS DRAMs, 158
novelty
complexity and, 2, 8
empirical setting and, 33
experimentation and, 22
innovation management and, 12
problem framing and, 108
problem-solving and, 106–107
in software, 174–175
software development and, 183–184, 185
study of technology integration and, 28–29
technology integration process and, 93
technology selection and, 2

Office Suite (Microsoft), 178
on-chip redundancy, 157
on-line environment, 173–175
on-line lead, role of, 177
Opus project, 178–181
organizations
communication issues, 69
development projects and, 14

organizations *(continued)*
performance and, 28–30
problem-solving and, 102–103, 107, 115–119
response to technological change, 121–123
structure, 14, 75–76
system-focused, 116–120
technological potential values, 124–126
technological yield values, 124–126
technology integration and, 4–5, 214–215
technology integration index and, 116–119

performance
measures, 82–83
organizational process and, 28–30
problem-solving process and, 31–32
project outcome and, 30–31
upper bound, 83
variation in, 35–36
See also product performance; project performance
person-years. *See* project resources (person-years)
Ph.D.-level hires, semiconductor industry, 166–167
plug and play, 3
plug-ins, 188
polyimide material technology, 129–130
potential. *See* technological potential
print-based lead, role of, 177
proactive technology integration, 3–5
problem framing, 107–108
problem-solving, 99–120
approaches to, 66, 101–102, 115–119
in bring-up stage, 103
frames of reference, 107–108
knowledge creation and, 99
knowledge domains and, 110, 114
mainframes, 100–106, 110–119
modeling as cyclical activity, 110
in novel and complex environments, 106–107
obvious solutions, 106
organizational process and, 102–103, 107, 115–119
path analysis, 108–110
performance and, 31–32, 114–115
robustness of early solutions, 118–119
server development, 103–106
solution generation, 107
technical content and, 111–112
workarounds, 104–105
problem-solving breadth, 106–107
context-independent, 110, 112
context-specific, 107, 110, 112, 114
product performance and, 111, 114–115, 119
project performance and, 110–114, 119
system focus and, 117t
technological yield and, 117t
process development
difficulties, 155
experimentation and, 156, 160
technology choice and, 158–159
technology integration capability and, 154–161
variation in, 35–36
process equipment, technology integration capability and, 164–165
processors
empirical approach to product performance, 84–85
performance measurement, 217–218
technical analysis, 217–221
technology choice, 84–87
process performance, 64
process yield, 156–157
product concept, technological choice and, 8–9
product development
domain-specific knowledge and, 10
flexible vs. traditional, 183–187
product integration, 75, 79
product manager, role of, 177
product performance, 75–96
assessment of, 30–31
defined, 83
determinants of, 76–79
development process and, 93–94
interconnect system, 220–221
measuring, 82–84
problem-solving and, 114–115

problem-solving breadth and, 111, 114–115, 119
process characteristics and, 88–94
product characteristics and, 94–95
project performance and, 87–89
research process and, 90–91
technology integration and, 28, 91–93, 99
variables, 87–89
See also performance; project performance
products
comparison indicators, 79–82
complex, subsystem interactions, 78–79
knowledge reflected in, 9–10
nature of, organizational structure and, 75–76
programmable memories, 155–159
program manager
role of, 177, 182–183
in software development, 178, 180–181, 182–183
project content, 38t
project performance and, 40–41
project experience, 61
deep, semiconductor industry, 168
in evolutionary projects, 143
See also experience
project lead, role of, 177
project level process, 28–30
project management, product performance and, 93–94
project organization, technology integration capability and, 163
project performance, 35–72
development process and, 42–47
internal integration and, 43–45
knowledge application and, 62–64, 67–69
knowledge generation and, 64, 69–70
knowledge retention and, 64–65, 70–71
lead-time variables, 38t, 39–42
measures of, 27, 28–29
person-years (project resources) and, 38–39
problem-solving breadth and, 110–114, 119
product outcome and, 87–89
project content and, 40–41
semiconductors, 56–59, 60–61
technology integration and, 28, 36, 47–51, 60–61, 65–71, 99
technology integration capability and, 50t, 51, 170, 171
traditional factors and, 46–47
variables affecting, 42
workstations and servers, 65–71
See also performance; product performance
project resources (person-years)
defined, 41t
development speed and, 42
effects of, 87–88
performance and, 38–39, 50t, 52t–54t
problem-solving breadth and, 111–112
technology integration, 52t–54t, 163–164
project team integration, performance and, 93
prototypes
My Yahoo!, 204
NetDynamics 2.0, 196, 204
Netscape Navigator, 188–190, 204–205
Silicon Graphics Challenge project, 69–70
software, 182
supercomputer packaging, 131
use of, 69–70

regression analysis, 222–223
research
institutions, 71
knowledge domain change and, 12–13
process characteristics, 90
product performance and, 90–91
technological yield and, 95
technology integration and, 21
technology integration capability and, 163–165
See also research experience
research and development (R&D)
challenges, 209–215
characteristics, 89–90
development projects, 13–14
inertia and, 15–18

research and development (R&D) *(continued)*
process effectiveness, 10
project performance and, 46
research projects, 12–13
technology change and, 15–16
technology integration needs, 211–212
technology transfer, 14–15
traditional, 4, 12–15, 46
uncertainty and, 205
research experience, 61
in revolutionary projects, 141–142
semiconductor industry, 166–167
See also experience; research
responsiveness, in software development, 183–184, 187, 205–208
revolutionary change
defined, 123
as foundation for evolutionary projects, 129–132
as response to technological change, 122–123
technological potential as measure of, 123–124
revolutionary projects
Challenge server, 134–138
characteristics of, 141–143
defined, 123–124
effectiveness of, 141–143
experimentation and, 141–142
inertia and, 145
knowledge generation in, 141–142, 145
massive experimentation in, 140
model for managing, 126–128
performance, regression analysis, 222
research experience and, 141–142
semiconductor industry, 169–170
in semiconductors, 124–126
variables associated with, 127t
rigidity, technological change and, 121
routinization, inertia and, 16–17
Runtime, 196–197

Samsung, 151
scalability, in Internet software development, 192, 194
science-based firms, challenges for, 200–215
scientific knowledge, 210–211
scientific method, 13
scripting language, selection of, 195–196
search engines, World Wide Web, 190–194
second-loop (second-order) learning process, 145
Semiconductor Research and Development Center (SRDC), 126
semiconductors, 51–65
empirical environment, 33, 55–56
evolutionary projects, 124–126, 169–170
experience, 166–169
experimentation in, 161–163, 167–168, 169–170, 212
focus in, 168–170
industry recovery, 150–165
Japan, 152–153, 165–170
knowledge application, 62–64, 168–169
knowledge generation, 64, 167–169
knowledge retention, 62t, 64–65, 166–167
Korean industry, 165–170
manufacturing capacity, 151–152
process assessment, 59–60
process technology, 55, 59–60
project performance, 52t, 56–59, 60–61
project turnover, 166
rapid technological change in, 55
research, 164–166
revolutionary projects, 124–126, 169–170
technology integration, 52t, 60–61, 212, 213
United States industry, 154–170
servers
integration and performance, 65–71
problem-solving, 103–106
SGI, IRIS 4d/480, 77–78, 79
shallow trench isolation, 125
"shooting the rapids," 185
Silbey, Alex, 140
Silicon Graphics, 184–185

Challenge project, 66–71, 134–138, 139
knowledge application, 67–69
Lego project, 18, 103–106, 134, 138–141
management of technological change by, 18, 19–20
technology integration process, 212, 214, 215
simulations, use of, 69
soft release (launch), software, 194, 203
software development
beta testing, 199–200, 202–203
complexity in, 176–177
concept development, 184–187, 201, 203–204
customer feedback in, 182
debugging, 179–180, 197–198
early development, 176
empirical setting, 33
external testing, 191
flexibility in, 183–187, 190–191, 205
implementation, 184–187, 201, 203–204
Internet, 187–201
knowledge development in, 210–213
leadership roles, 177
Microsoft Explorer, 199–201
models, 184
Netscape Navigator, 187–190, 201
novelty and complexity in, 174–175
processes, 187–208
prototypes, 182
responsiveness in, 183–184, 187, 205–208
scalability in, 192, 194
slow release, 191
soft launch, 194
technology integration for, 174–177, 213
traditional, 183–187
Word for Windows, 177–181
Yahoo!, 190–194, 201
See also Internet software
specialization, deep, in research projects, 13
SPECmark, 77
speed
in software development, 185
workstations and server requirements, 66
SS180, 79–81, 85
stabilization, of code, 179
Stanford University
Dash project, 138, 139
Electrical Engineering Department, 71
strategy, technology integration process and, 213–215
Studio, 196–197
subsystem interactions, in complex products, 78–79
subsystem-level models, 131
Sun Microsystems, 196
supercomputers
empirical approach to product performance, 84–85
experimentation, 212
packaging, 130–131
technological evolution, 128–134
suppliers, involvement of, 159–160
SX supercomputer series, 128–134
system-focused organizations
holistic view of, 119–120
technology integration index, 116–119
system-focused technology development, 95
system knowledge, defined, 11, 24

TEA laser, 11
technical content, 38t
defined, 40–41
problem-solving and, 111–112
technical lead, 175–176, 177
technological base, stability of, 2
technological change
evolutionary response to, 122–123
inertia and, 16–17, 121
managing, 18–20, 121–146
organizational response to, 121–123
rapid, 17, 55
recognition of need for, 17
research and development organizations and, 15–16
revolutionary response to, 122–123

technological change *(continued)*
rigidity and, 121
semiconductors, 55
technology integration process and, 145–146
See also evolutionary change; revolutionary change
technological evolution. *See* evolutionary change; evolutionary projects
technological potential
advanced computer processors, 84–85
calculating, for interconnect system, 219
context-specific breadth in problem-solving and, 114–115
defined, 31, 82, 83
development process and, 94t
experimentation and, 93
factors affecting, 95
knowledge domains and, 89
as measure of revolution, 123–124
measuring, 82–84, 85–87
in microprocessor organizations, 124–126
project resources and, 88
realized, 85–87
regression analysis, 222
research process and, 90, 91t
technology integration process and, 91, 92t
workstations, 82
technological revolution. *See* revolutionary change; revolutionary projects
technological stability, inertia and, 16
technological yield
defined, 31, 82
development process and, 94t
domain interactions and, 89
factors affecting, 94–95
market introduction and, 95
as measure of evolution, 123–124
measuring, 82–84, 85–87
in microprocessor organizations, 124–126
problem-solving breadth and, 114–115, 117t
processor module technical analysis, 217–221
product performance and, 94–95
project resources and, 88
regression analysis, 222
research and, 90–91, 95
targeted technology integration process and, 89
technology integration and, 89, 91–93, 95
workstations, 82
technology choice
in advanced processors, 84–87
analyzing, 82–84
approaches to, 36
assessment of, 31
complexity and, 71–72
context-specific knowledge and, 114
decisions, 8
experience and, 168
experimentation and, 64, 161
knowledge generation and, 21–22
management and, 20
performance variables, 77–79
proactive, integrative process for, 20–21
problems, 8
process development and, 156, 158–159
product systems, 79–82
research and development and, 211
scripting language, 195–196
technological potential and, 95
unconscious, 20
under uncertainty, 71–72
technology integration
applications, 5
assessment of differences in, 29–30
conceptual framework for, 5, 9–25
defined, 1, 8, 21
effectiveness of, 4–5, 24–25, 213–215
failures of, 180–181
flexibility and, 183–187
foundations of, 102–103
importance of, 1–2, 20–21, 63
improving process of, 212–213
Internet software, 174–177, 206–208
knowledge and, 21
for management of technological change, 18, 121, 145–146

need for, 211–212
organizational processes and, 4–5
proactive, 3–5
process assessment, 59–60
product performance and, 91–93, 99
program manager role in, 182
project performance and, 36, 47–51, 60–61, 65–71, 99
role of, 20–21
roots of, 119–120
semiconductors, 59–61, 63
strategic impacts of, 213–215
technological yield and, 89, 95
technology selection and, 20–21
validation through, 214
Word for Windows, 180–181
technology integration capability, 149–171
flexible product development model and, 187
process development and, 154–161
project performance and, 170, 171
in semiconductor industry, 154
in software industries, 175
technology integration capability index, 46–47
knowledge application and, 47t
knowledge generation and, 48t
knowledge retention and, 49t
project performance and, 50t, 51
technology integration index
by geographic area, 52t, 53t
organizational process and, 116–119
technology selection. *See* technology choice
technology transfer
"Copy Exactly," 160, 169
factors affecting, 14–15
testing
external, 70, 191, 202, 203
internal, 199, 202–203
user, 202–203
using intranets, 199–200
See also beta testing; feedback
Texas Instruments, 154, 212
DRAMs, 151
semiconductors, 150–151
theoretical maximum performance, 83
theoretical potential, 84
time scales, Internet, 174
time to market, in software development, 185
T. J. Watson Research Center, 125–126, 150, 209
Toshiba, 51, 55, 151
total lead time, 38t
defined, 40, 41t
effects of, 87–88
in software development, 185
transistor (gate) density. *See* gate density
turnover
in evolutionary projects, 144
performance and, 64–65
semiconductor industry, 166
technology integration failure and, 180

uncertainty
knowledge domains, 12
problem framing and, 108
research and development and, 205, 211
scientific knowledge and, 211
technical choice and, 71–72
United States
computer industry losses, 149
computer industry recovery, 149–165
semiconductor industry, 154–170
technology integration capability index, 49
technology integration/project performance relationship, 53t
universities, 71
unpredictability. *See* uncertainty
user testing
Internet software development, 201, 202–203
See also beta testing; feedback; testing
U-shaped curve, 157, 161

validation, through technology integration, 214
video on demand (VOD), 173

Watson, T. J., Research Laboratory, 125–126, 150, 209
Windows, 177, 180
Windows 95, 3

Windows NT, 194
Word for Windows 1.0, 177–181
  technology integration failures, 180–181
workarounds, problem-solving and, 104–105
workstations
  empirical setting, 33
  integration and performance, 65–71
  performance characteristics, 81–82
  product performance determinants, 77–78
  scientific knowledge base, 210–211
World Wide Web, Yahoo!, 190–194

Yahoo!, 173, 190–194, 203–204
  knowledge development and, 202
  My Yahoo!, 190–194, 203–204
  technology integration process, 214
yield. *See* technological yield
Yen, Wei, 140

"zero defects," 181
Z1000, 79–82, 84

About the Author

Marco Iansiti is an associate professor in technology and operations management at Harvard Business School. His research focuses on the management of technology and product development. He has been involved in numerous studies of effective development practices in many industries, ranging from microelectronics to steel manufacturing, and from software to automobiles. He recently completed a worldwide study of the methods and practices of technology development in the microelectronics and computer industries, comparing Japanese, U.S., European, and Korean companies. The focus of this study was to better understand the drivers of firm capability in developing products based on rapidly changing technologies. Professor Iansiti is currently involved in a new study that looks at product development in environments characterized by extreme market and technological turbulence. This new study focuses on software, multimedia, and workstations.

Professor Iansiti has worked as a consultant to several major Fortune 500 companies in industries ranging from medical products to telecommunications, and he is a board member of the Corporate Design Foundation.

Professor Iansiti has authored and coauthored more than three dozen papers, book chapters, articles, and cases. His work has appeared in a variety of journals, including the *Harvard Business Review, The California Management Review, Research Policy, Industrial and Corporate Change, Production and Operations Management*, and *IEEE Transactions on Engineering Management.*

Ask Marco:

1) Does Integration really work when the pure research team has no product responsibility?

2) Will the integration model agree with Henderson & Clark model of organizational innovation? Does integration conflict with organizational knowledge of the technology architecture

3) Iansiti's focus is primarily high tech. What about heavy industries? or services? especially in the context of von Hippel's "sticky information"?